Daffodil

Biography of a Flower

Helen O'Neill

HarperCollins*Publishers*

For my parents

HarperCollins*Publishers*
First published in Australia in 2016
by HarperCollins*Publishers* Australia Pty Limited
ABN 36 009 913 517
harpercollins.com.au

HarperCollins*Publishers*
Level 13, 201 Elizabeth Street, Sydney NSW 2000, Australia
Unit D1, 63 Apollo Drive, Rosedale, Auckland 0632, New Zealand
A 53, Sector 57, Noida, UP, India
1 London Bridge Street, London, SE1 9GF, United Kingdom
2 Bloor Street East, 20th floor, Toronto, Ontario M4W 1A8, Canada
195 Broadway, New York NY 10007, USA

National Library of Australia Cataloguing-in-Pubication data:

O'Neill, Helen, author.
 Daffodil : the biography of a flower / Helen O'Neill.
 ISBN: 9780732299200 (hardback)
 Subjects: Daffodils.
 Narcissus (Plants)
584.34

Cover image © Red Edge / Rebecca Cozart
Cover and internal design by Hazel Lam, HarperCollins Design Studio
Typeset in Hoefler Text by Jonathan Hoefler
Colour reproduction by Graphic Print Group, South Australia
Printed and bound in China by RR Donnelley on 128 gsm Huaxia Sun matte art paper
8 7 6 5 4 3 2 1   16 17 18 19 20

CONTENTS

'He that has two cakes of bread let him sell one of them for some flowers of the Narcissus, for bread is food for the body but Narcissus is food of the soul.'

attributed to Mohammed
(c. AD 570–632)

Daffodils have been part of my life for as long as I can remember. Born in Hampshire's New Forest my roots lie deep in the rich soil of southern England, yet my childhood was one of perpetual motion as my family moved from one home to another, prompted by advances in my father's career.

Across my ever-changing world daffodils became a constant. As each winter receded they appeared anew, a radiant signal that the bleakest English season was done with and the New Year truly on its way. By the time I hit my teens my family's travelling halted, and we settled in the countryside a few miles from a Thames Valley village. My new home was surrounded by towering woodland dissected by pathways that had been trod for centuries, dappled meadows carpeted all-too-briefly with bluebells — and each spring what felt like acres of drifting daffodils.

It was then I absorbed the fact, without ever having to think about it, that the daffodil is not one flower but many. Ours blossomed in a prismatic kaleidoscope of colours from tissue paper white to the deepest blood orange and in a melange of flower forms and sizes. Some were elfin, others giants brandishing flowers that ranged in shape from classic golden trumpets to creamy stars with twisting petals and tiny butter-lemon cups.

As one daffodil variety melted away another materialised to take its place, a rhythmic dance through the spring chill that lasted, it seemed to my young mind, for ages. The blossoms were beautiful, injecting a lifeblood of colour into the drained winter landscape and we took them for granted. After all, they were simply daffodils.

As an adult I transplanted myself to Sydney on Australia's dazzling east coast, a magical place where technicolour parrots wheel about like rambunctious toddlers, fruit bats are the size of pussycats and the indigenous foliage appeared — to my eye — alien indeed.

Yet cut bunches of daffodils appeared in flower shop displays early each spring and daffodils could be seen scattered across gardens in this arid continent's cooler regions. They, like me, could not be called native yet clearly felt at home, particularly in the mountains of Victoria and New South Wales, the high districts around Perth in Western Australia and in certain locations across Tasmania and the landlocked ACT.

I lived in Sydney for decades and acclimatised completely, or so I thought. In late 2008 the opportunity arose to move back to Europe for six months and take up an Australia Council literary residency in a Paris apartment called the Keesing Studio. My partner and I boxed up the contents of our Bondi home, packed our warmest clothes and moved into the cosy atelier on the Right Bank of the River Seine.

It was early February and Paris was at its most desolate, winter having drained every atom of colour from the city. Impulse drove me to a cramped garden supplies store where I bought window boxes, potting mixture and dozens of miniature daffodil bulbs. I planted the little zombies as deep as the window boxes would allow, fired up my computer and immersed myself in work.

Gradually the plants emerged from the chilled soil, their razor-sharp leaves slicing into the air before their stems budded and burst into bright yellow flowers that faced down the end of winter and danced with the spring breeze. These daffodils blossomed for almost a month, ushered in the warmer weather and then, just as stealthily, faded away.

At the end of my Parisian stay I cleared out my window boxes, brushed the soil from the bulbs and sealed them inside a large, white envelope which I stowed away into a drawer. That is where Sophie Masson, a French-Australian children's author, found them at the beginning of her winter Keesing Studio stay. Delighted, she told me she immediately replanted the bulbs. They made her feel intimately connected with the frozen city and filled her with reassurance that whatever else might happen, spring, and the blossoms, would come.

Later that same year back in Sydney I received a shock diagnosis of breast cancer. Emotionally my world froze in the face of one of the most witheringly hot summers I had ever experienced. I underwent surgeries and embarked on long rounds of chemotherapy, radiotherapy and finally hormone treatments. As my cancer-fighting regime progressed and I became weaker, in the northern hemisphere winter gave way to spring. Across my parents' gardens the daffodil army mustered and bloomed.

My brother started photographing the daffodils and sending me his glorious pictures. His intention was to convey a message of hope, to help me realise the hard times would pass and that life would again be bright. He was not the only person to use the daffodil as an emissary. A friend in Melbourne who had herself beaten

cancer posted me a sweet package of soft, cotton headscarves with a note, written in a card decorated by a lovely line-drawn image of *Narcissus poeticus,* letting me know I could ring any time. Another well-wisher from America sent me words of cheer and a beautiful, pink, stylised daffodil pin.

My brother was right, the hard times did pass, and as I began to recover I started thinking about the daffodil.

My parents had inherited a remarkable landscape where, each spring, a multitude of daffodils bloomed in flashes, clumps and languid drifts. My mother was entranced. On a whim she severed the stems of some particularly fine specimens and entered them in the village show. She returned home with prize certificates for best blooms and a prestigious silver cup. The following year she entered another batch. The same thing happened again.

My mother had by now picked up some tricks when it came to exhibiting. She found that the best times to pick her flowers were early in the morning or late in the afternoon. Whether she used a knife to cut the stems or flexed them hard until they snapped, she held the broken ends uppermost in her hand to stop the sap running out. She kept a bucket of water handy and popped each daffodil in as quickly as she could. Critically, she honed her sense of which blooms to select, how to sidestep pre-show stem and petal stress, and when to fire up her hairdryer in order to coax stubborn buds into opening in time for their big moment.

Curiosity gnawed away at her. Of what did her daffodil menagerie really comprise? She began cataloguing them, drawing reference sketches of what seemed to be dozens of handsome, olde worlde varieties, and even invited a 'bulb hunter' to visit, a horticulturalist who sporadically spent time touring out-of-the-way private gardens in the hope that they might contain some vintage cultivars his commercial collection lacked.

I happened to be present on the clear spring morning in 2006 when the plant expert studiously inventoried my mother's daffodils, identifying the different heritage varieties that happened to be in bloom that day with vintage names such as 'Victoria', 'Empress', 'Bath's Flame', 'Sunrise' and 'Star'. In reply to questions about their origins he dropped tantalising snippets of information about long-dead breeders, alluring titbits about a mysterious, apparently ancient daffodil world.

That visit proved a turning point. I started jotting down little pieces of daffodil wisdom in a notepad to be checked at a later date. My list included 'facts' such as:

- The words 'daffodil' and '*Narcissus*' are often used interchangeably, but daffodils (a group that includes jonquils) belong to the botanical genus *Narcissus*. The full classification is as follows:

    Domain — Eukarya;

    Kingdom — Plantae;

    Phylum — Magnoliophyta;

    Class — Liliopsida;

    Order — Liliales;

    Family — Amaryllidaceae;

    Genus — *Narcissus*.

- Leeks, chives, onions, garlic, amaryllis, snowdrop and the Amazon lily also belong to the Amaryllidaceae family.

- Over 30,000 different daffodil cultivars have been bred into existence by hybridisers. Just a small fraction of the flowers created survive to this day.

- Botanists recognise fifty-four *Narcissus* species plus naturally occurring hybrids, yet a 2008 DNA investigation determined the total number of species was thirty-six.

- Daffodils are native to England/Great Britain.
- Daffodils are *not* native to England/Great Britain.
- 'Tête-à-Tête', one of the most popular and famous miniature varieties is sterile. Every bulb is a clone of the original plant, derived from a cross that occurred entirely by accident. The identity of its parents is unknown.
- The daffodil has been used as a form of currency. The Duchy of Cornwall is paid a peppercorn rent of a single daffodil each year by the Isles of Scilly Wildlife Trust.
- Ramses II, Ancient Egypt's great pharaoh, supposedly had daffodil bulbs placed on his eyes after death.
- One man, called Peter Barr, purportedly created the 'modern' daffodil. He is known to *Narcissus* cognoscenti as the 'Daffodil King'.

Every now and again I wondered whether daffodils had started to haunt me. Years ago I wrote a biography about Florence Broadhurst, the murdered Australian wallpaper designer responsible for hundreds of boldly iconic, highly recognisable hand-drawn patterns. Some Florence Broadhurst images have long decorated my office walls, including a detail from one rather lyrical floral image entitled *Carnation* which I suddenly found myself gazing at properly for the very first time.

Wanting to be certain of what I thought I could see I visited Signature Prints, the Sydney company that prints Florence Broadhurst designs, and asked its CEO, David Lennie, if he was aware that *Carnation*, one of the designer's best-known patterns, was actually full of daffodils — jonquils, to be precise. He bluntly told me if it was he had never seen them, so we found the large silk screen containing the image in question. I pointed to the errant blooms. Lennie peered at them and simply said, 'Oh.'

While I was making arcane lists and collecting daffodil ephemera (thank you, eBay) my mother was busy bolstering her living daffodil collection by buying bulbs from Cornish horticultural company R.A. Scamp, whose catalogues proved a revelation. Full of useful reference images they offered a multitude of cultivars identified by 'Division' (of which there are thirteen — see page 224 for the full Royal Horticultural Society list), colour code (W for white, Y for yellow and so on), bloom date (from V.E. for 'Very early' to V.L. meaning 'Very late') and — of course — name.

The daffodils' monikers as listed in the catalogues made a beautiful, inventive, deliciously varied inventory, employing such names as 'Lucifer', 'Intrigue', 'Hanky Panky', 'Hot Gossip', 'Odd Job', 'Cornish Chuckles', 'Madame Speaker', 'Flirt', 'Foundling', 'Dragon Run' and 'Helford Dawn'.

Confronted with prices that bolted up to £25 a bulb my mother opted for a relatively modest assortment of heritage cultivars, added some hot-off-the-press 'new' breeds for good measure, and topped off her choice with something sold as 'Ronnie's Rainbow Mixture', a lucky dip of 100 unidentified daffodil bulbs that seemed a bargain at just £10.

My mother did not know it but Ron Scamp, the man who founded this firm, is something of a celebrity in the daffodil world. He prides himself on breeding more daffodil varieties than any other grower, telling me when I called him up to find out more that he supplied over 3,000 varieties of daffodil, an agglomeration that represented, as he put it, 'a lifetime's collecting and growing'.

As we chatted away Scamp ticked off the destinations of his far-flung offspring: 'Australia, New Zealand, South America, North America ... Canada ... all countries in Europe, we've had some taken to India, [to] the Middle East, where certain of the daffodils that like the hotter type of climate are grown [and] a lot into Japan.'

You might even claim that daffodils were in his blood, he said, only half-jokingly. Scamp had spent part of his childhood living on a Cornish flower farm belonging to his uncle Dan du Plessis (1924–2001), a connoisseur who hybridised avant-garde strains and won some serious *Narcissus* awards. Scamp's own interest was seeded as a boy.

'I used to pick snowdrops, and lent lilies, the wild daffodils and primroses and send them to Covent Garden market ... for pocket money,' he said, describing how he pursued a career in retail management until daffodils lured him away. Like his forebear, he became gripped by the thrill of experimenting. He bred the little creatures initially with his uncle Dan's interests in mind. 'He wanted daffodils that extended the season, daffodils that were better quality doubles, new types of shapes and colours,' explained Scamp, adding that on a personal level, 'Daffodils became a hobby, then a passion, then an obsession.'

When Dan retired, Scamp took over the business. In 1993 he named a new long-cupped yellow trumpet 'Ouma', meaning 'great mother' in Afrikaans, in honour of his grandmother. The following year his mother pointedly informed him that he had grown 'a meadow full of new varieties' yet never christened one after her.

'I said, "Come on we'll look," and we found a daffodil in the field,' he said of a flat-faced, split-cupped collar *Narcissus* with wavy margins, a yellow base, and a deep orange-red corona (trumpet or cup). 'We named it "Max" because she was affectionately known as Max in the family.' Another of his creations — a multi-headed, yellow and orange tazetta daffodil — he titled 'Dan du Plessis' in honour of his pioneering uncle.

Scamp has named flowers after everything from Cornish tin mines to stately homes. Other breeders have gone further still. 'For love, for family, for memories, even after dogs — every named daffodil has a story,' Scamp told me. 'Behind every one is a tale.'

The more I delved into the daffodil the truer Scamp's words rang. This flower has implanted itself into virtually every cultural recess imaginable, from psychology and poetry to popular culture. Daffodils even featured in the cult television series *Doctor Who*, thanks to the 'Terror of the Autons' storyline in which the Master deviously tried to destroy the entire human race using deadly yellow blooms created from living plastic.

Yet the intimate story of the daffodil is as outlandish as any science fiction fable. This flower's sex life has startled scientists, as have some of the most fundamental facts about this beautiful, multi-faceted plant. The image it portrays is that of a quintessentially British creature, yet the English shunned it for centuries until an unlikely league of nineteenth-century obsessives re-engineered its future. Today the daffodil is arguably the world's most powerful flower, and its tale is one of passion, influence, mythmaking and romance.

CHAPTER I

Dawn of the Daffodil

'The Daffodil, as grown in our
gardens, is a purely English flower.'

Reverend George Herbert Engleheart,

Introduction to *Daffodil Growing for Pleasure and Profit* by Albert F. Calvert, 1929

A treacherous mist rolled through England's Lake District on the morning of Thursday, 15 April 1802, and a storm seemed on its way but Dorothy Wordsworth and her older brother William decided to start their countryside trek anyway. As is clear from Dorothy's private journal the siblings had spent the previous night at a friend's estate near the Cumbrian village of Pooley Bridge and the time had come to walk back to Dove Cottage, their home in Grasmere.

As the Wordsworths hiked, the wind grew so fierce they took refuge under a furze bush and briefly considered turning back. Instead they ploughed on, tramping through fields and woodland beside Ullswater Lake, where they came across a spectacle that took Dorothy's breath quite away. At first just a few daffodils appeared here and there, blooming close to the water's edge. Dorothy never described these in detail but there is little doubt they were *Narcissus*

DAFFODIL

20

pseudonarcissus, a delicate, golden-tinged trumpet daffodil; one of the most ancient varieties of the flower:

> But as we went along there were more & yet more & at last under the boughs of the trees ... a long belt of them along the shore, about the breadth of a country turnpike road. I never saw daffodils so beautiful they grew among the mossy stones about & about them, some rested their heads upon these stones as on a pillow for weariness & the rest tossed & reeled & danced & seemed as if they verily laughed with the wind that blew upon them over the Lake, they looked so gay ever glancing ever changing.

Two years later Dorothy's brother metaphorically revisited that moment and, using poetic licence in its purest form, edited his sister out of the picture. He wrote a series of verses beginning with the words:

> I wandered lonely as a Cloud
> That floats on high o'er Vales and Hills,
> When all at once I saw a crowd,
> A host of dancing Daffodils ...

William Wordsworth told his ode from the point of view of a desolate man stunned by a glorious battalion of flowers who would only much later realise the significance of what he had seen as the memory flashed 'upon that inward eye':

> Which is the bliss of solitude;
> And then my heart with pleasure fills,
> And dances with the Daffodils.

The creation of the poem really does seem to have been a family affair, the poet even crediting his wife Mary with some of the lines. Titled 'I Wandered Lonely as a Cloud', it was published in Wordsworth's 1807 work *Poems, in Two Volumes* within which another verse, 'The Sparrow's Nest', paid tribute to his sister Dorothy with the words: 'She gave me eyes, she gave me ears.' His collection was savaged as 'trash ... an insult on the public taste' by the literary critic Francis Jeffrey, who did not even bother mentioning 'I Wandered Lonely as a Cloud'. The poet continued working at it and on 1815 delivered the definitive, 24-line version of what would become one of the best-known poems ever written in the English language.

When William Wordsworth died from pleurisy in 1850 he was regarded as one of England's most significant writers, yet the daffodil that had inspired his greatest work was considered by many as, at best, just another weed. Both gardeners and the public largely shunned it, as they had for centuries, and understanding why involves drifting back in time.

Just where and when the first daffodil bloomed remains something of a mystery, but one point upon which researchers agree is that it was certainly nowhere near Britain. Many hold that the flower originated in south-west Europe's Iberian Peninsula, an area encompassing countries such as Portugal, Spain and France that represents a hotspot of *Narcissus* diversity. The first daffodil appeared at some point between around 29 and 18 million years ago (the late Oligocene and early Miocine eras), making it what Spencer Barrett, Professor of Ecology and Evolutionary Biology at the University of Toronto, terms a 'relative newcomer to the story' of evolution. It is, he told me, 'quite an advanced flowering plant'. Humans, of course, are newer still. For eons we existed as hunter-gatherers until around 10,000 years ago we began experimenting with domesticating the wild creatures around us.

This evolutionary milestone would ultimately prompt reliance on a tiny fraction of Earth's plants and animals, and result in humans becoming arguably the most powerful single species on Earth.

Unlike wheat, cotton, rice or maize the daffodil appeared to have little obvious use. Ancient cultures feared it with good reason — its bulbs, stems, leaves and flowers are all toxic — and some civilisations saw it as a living link to the hereafter, an element in myths that sought to unravel the mysteries of life after death. A fraught relationship developed between daffodils and humans but as we explored, traded, invaded and subjugated each other it would appear that we took *Narcissus* along for the ride.

In around 300 BC the Ancient Greek philosopher and botanist Theophrastus listed various daffodil strains in his nine-volume magnum opus *Enquiry into Plants*. He described the plant's structure, remarked that garland-makers prized it and implied that his contemporaries even cultivated it. Theophrastus used the word 'nardissos' to describe the daffodil, a term the Romans would transmute into 'narcissus', and two theories compete as to why his culture selected this term.

The first theory relates to the Greek word *narkao* (ναρκάω), the root of the term 'narcotic', and the associated belief that daffodils, when eaten, had hazardously stupefying properties or, as various legends implied, quasi-demonic overtones. One legend has it that daffodils lured Persephone, the beauteous daughter of the gods Demeter and Zeus, to the Underworld. As Persephone bent down to pluck some daffodils the earth opened and the god Hades swept her into his realm.

The second theory draws from the myth of Narcissus, a tale that ripples still through art, iconography, drama, psychology and popular culture. It spawned the notion of narcissism, refashioned by Sigmund Freud into a pervasive psychoanalytic concept that resonates today in the 21st-century's selfie-obsessed digital landscape. The Roman poet Ovid published the best-known telling of the

Narcissus story in AD 8 almost as an aside within his multivolume epic tome *The Metamorphoses*.

Ovid's Narcissus is the son of a nymph called Liriope and the river god Cephissus. He is devastatingly attractive, admired by males and females alike, and so magnetic that a blind prophet called Tiresias warns Narcissus's mother that he will only grow old if he never sees his own face.

Narcissus blithely breaks the heart of Echo, a nymph who pines away, able only to repeat his words, until she is nothing more than a ghostly disembodied voice. The youth's scornful behaviour so appals the gods they decree that he shall never find satisfaction in love. One day while out hunting, Narcissus catches sight of his own reflection in water and falls desperately in love. Transfixed, he refuses to leave the object of his desire and wastes slowly away to death at the water's edge. As nymphs mourn his tragic passing a flower grows from the soil, the head of which droops as though it, too, is gazing into its own reflection. The Greeks named the flower after the pitiful young man: *Narcissus.*

Some of Western culture's most influential artists and thinkers drew inspiration from Ovid's telling of this myth, yet in 2004, almost 2,000 years after Ovid's publication, a British scholar announced the unearthing of an earlier, more harrowing version of the tale.

Dr Benjamin Henry, then of Oxford University's classics faculty, had been examining fragments of papyrus originally salvaged from an ancient rubbish tip near the Egyptian settlement of Oxyrhynchus. Tens of thousands of such scraps had been retrieved by British archaeological adventurers during the late nineteenth and early twentieth centuries, saved and transported to places such as Oxford University where they remained, metaphorically gathering dust.

As Henry deciphered a particular fragment he realised that it held a blunter, uglier rendition of Ovid's Narcissus legend, which appeared to have been penned half a century earlier. The author, Henry suggested, could have been Parthenius

of Nicaea, a first-century BC Greek scholar taken prisoner by the Romans, who is believed to have tutored Virgil, and of whom Ovid would have undoubtedly known.

To this day the 'Narcissus fragment' remains arguably the myth's earliest incarnation (not all Ovid scholars agree with Henry's findings) and it is, even by 21st-century sensibilities, violent indeed. In this version the love-struck Narcissus does not waste poignantly away by the water's edge. Instead, the impact of glimpsing his reflection is instantaneous and catastrophic. The youth wails, weeps for his beauty and reacts with a maniacal savagery — attacking his own body so ferociously with a sharp blade that he kills himself on the spot.

The shell-shocked youth's blood spilled on to the ground and from it the *Narcissus* flower grew. Henry, now a Research Associate at University College London's Department of Latin and Greek, speculates that Ovid was familiar with this earlier version and deliberately censored it, gutting its wild horror to create a more palatable tale. In so doing Ovid fundamentally altered the story, and with that our understanding of what the Greeks really thought of the daffodil, a flower borne not from the poetically tragic image of a beautiful youth pining gently away but from a maelstrom of violent suicide. What powerful cultural shapers such as Shakespeare, Milton, Goethe, Rousseau, Salvador Dali and Sigmund Freud would have made of Narcissus's hysteria we can never know.

Certainly the Romans, like the Greeks, had a healthy respect for the daffodil. One often repeated 'fact' is that the essential kit of some Roman soldiers may have included *Narcissus* bulbs which could be employed as suicide pills in the event of capture or catastrophic injury. Even so, death by daffodil appears to have been a relatively unusual affair. The Roman Emperor Commodus died in his bathtub in the year AD 192 at the hands of an elite wrestler-turned-assassin called Narcissus (upon whom Russell Crowe's character in the Ridley Scott film *Gladiator* was partially based) but that does not really count.

The notion that *Narcissus pseudonarcissus* hitched a lift with the Romans to

Britain, where it escaped and grows feral to this day, sounds plausible but there may be more to that tale. Horticulturalist Richard Wilford, Manager of Garden Design and Collection Support at Kew's Royal Botanic Gardens, will not rule out the possibility that England may actually represent the boundary of the daffodil's original natural range. 'Many species were pushed back from the British Isles into mainland Europe by the Ice Age,' he told me. 'It can be difficult to know if they re-colonised naturally once the ice retreated or if they were introduced by man.'

Either way *Narcissus pseudonarcissus* colonised England and Wales as well as various countries of mainland Europe. In various places it spawned different names, as Jan Dalton, the British Daffodil Society's Archivist, realised in 2011 upon unearthing 125 regional monikers for the daffodil in Britain alone, most of which appeared to refer to *Narcissus pseudonarcissus*. Dalton's tally included terms such as English Lent Lily, Lide Lily, Solomon's Lily, Yellow Lily, Bell Rose, Corn Flower, Cowslip, Crow Foot, Giggary, Glens, Gracy Day, Haverdrils, Julians, Lent Cocks and Wyrt.

All of which begs the question of how the word 'daffodil' itself came to be. As usual there are various theories: that it stems from Latin (which has two words: *affodilus* and *asphodilus*), from Middle English (*affodill*) or from the Dutch (*de affodil*). Whatever the source, the English had some fun with the term. *Narcissus pseudonarcissus* became known as affodylle, affadile, asphodel, daffadowndilly, daffodilly, affodilly and, of course, daffodil.

Whatever people chose to call it, this plant took its place on the brutal roulette wheel of herbal medicine during the Middle Ages. Reference to daffodils appears in the writings of John Arderne (1307–1376), a mysterious British herbalist and physician now regarded by some as one of the founding fathers of surgery. Though born into a barbaric time transfixed by religious superstition, Arderne attempted to logically analyse his patients' ailments and record what he learned, using drawings of everything from flora to corpse anatomy and, for his era, an admirably scientific approach.

The British Library holds an illuminated Arderne manuscript within which appears a line drawing of an 'Affadille'; a recognisable daffodil. His textbook on surgery, *De Arte Phisicali et de Circurgia*, recommends the toxic flower as a 'medicinal' plant in a rather dubious-sounding concoction, a recipe involving an oil and wine blend 'decocted with centaury and [daffodil] roots', aimed at combating 'cold inflammation'.

Arderne understood that art exhibiting some degree of accuracy could be a powerful tool in disseminating information, however dangerous that 'knowledge' might be. The advent of German inventor Johannes Gutenberg's printing press around twenty-five years later revolutionised Renaissance Europe by paving the way for mass communication — a pathway for knowledge, and ideas, to spread. In 1530 German theologian, botanist and physician Otto Brunfels published the inaugural instalment of what would be the world's first botanical bestseller, his encyclopaedic three-part catalogue *Herbarum vivae eicones ad naturae imitationem* (Portraits of Living Plants). It featured illustrations by Hans Weiditz, a pupil of Albrecht Dürer and one of the most skilled woodcutters — and under-appreciated artists — ever to exist.

Weiditz created thoughtful, detailed and painstakingly accurate woodcut prints that allowed any reader, anywhere, at any time of year to potentially identify without question the plants Brunfels described. Weiditz's work ushered in the practice of accurate botanical artistry and, beyond this, a radically novel approach that separated the woolly idea of 'plant knowledge' from the precise framework of medicine. He paved the way for scientific botany to become an area of investigation in its own right.

Weiditz depicted two daffodil species — *Narcissus pseudonarcissus* and *Narcissus nobilis* (a wild *Narcissus pseudonarcissus* variant with light petals and a bold, long trumpet stained a deep yellow) — with such accuracy that over 470 years later when a group of researchers decided to track *Narcissus pseudonarcissus*'s journey through time and space they realised they could begin with his prints.

Narcissus pseudonarcissus is not just any species. Occasionally known as the 'Ajax group' it is one of the most significant ancestors of all contemporary daffodils, and it is argued by some to be the parent of up to 99 per cent of yellow trumpet cultivars. It travelled far and wide from its original Mediterranean home to Holland, France and English gardens like those I grew up enjoying, but nobody really understood when, why or how.

Botany professor Diego Rivera from the Department of Plant Biology at Spain's University of Murcia, one of the scientists involved in the research project, told me they used art and literature to track *N. pseudonarcissus* through eleventh-century Arabia and fifteenth-century Flanders, where weavers worked intricate *Narcissus* flowers into rugs depicting myths and Bible tales. From the outset this daffodil's timeline appeared intrinsically confusing. The living populations are scattered. 'Most of them are feral, so escaped from very ancient cultivation,' Rivera explained. 'Even some of the Spanish populations could be feral, because daffodils were in cultivation in Islamic gardens in medieval times.'

During the Renaissance, for the first time in human history, art turned into a form of science thanks to the exquisite observational accuracy of fifteenth-century practitioners such as Albrecht Dürer, Michelangelo and Leonardo da Vinci. Weiditz's art proved pivotal to both the wider movement, and the daffodil researchers' hunt. The resulting paper, 'The Image of Daffodil in Art and Botanical Illustration: Clues to the History of Domestication and Selection of *Narcissus* subgenus *Ajax* (Amaryllidaceae)', attempted to build a four-dimensional roadmap of *pseudonarcissus* and generate a family tree of twenty-six wild and cultivated daffodil groups.

Daffodils appeared to have been domesticated on at least three separate occasions: during the Middle Ages, in the sixteenth and seventeenth centuries, and then again in the 1800s. The researchers tracked the flower through Renaissance Europe, noting that in England daffodils were largely ignored until the mid-1500s, and singled out several individuals whose actions had sparked change, including

PARADISI IN SOLE
Paradisus Terrestris.
or
A Garden of all sorts of pleasant flowers which our
English ayre will permitt to be noursed vp:
with
A Kitchen garden of all manner of herbes, rootes, & fruites,
for meate or sause vsed with vs,
and
An Orchard of all sorte of fruitbearing Trees

a sixteenth-century man, the son of a French gardener called John Robin, whose name is lost in time but who seems to have been responsible for importing various strains of daffodil from Spain into French gardens.

Special mention goes to a seventeenth-century English botanist whose intellectual brilliance saw him breach the chasm between superstition and science. Despite this man's achievements he would be neglected by history, just as the daffodil appeared to have been. His name was John Parkinson, and the scientific paper does not tell his full tale.

In 1567, the year Mary Queen of Scots' forced abdication resulted in her one-year-old son James being crowned King James VI of Scotland, John Parkinson was born to an apparently unremarkable farming family in Lancashire, northern England.

Ambitious and determined, Parkinson left for London at the age of fourteen with the hope of training as an apothecary. He emerged as one of the most celebrated and powerful herbalists of his day, becoming Royal Apothecary to King James I of England, and after James's death in 1625, a favourite of the new monarch, Charles I.

In 1629 Parkinson published an illustrated book called *Paradisi in Sole Paradisus Terrestris* with a flourish of sardonic self-confidence and some mischievous Latin–English wordplay: its title translates rather cheekily to 'Park-in-Sun's Earthly Paradise'.

The book was dedicated to Charles I's young wife, Henrietta Maria of France, and its publication resulted in Parkinson being awarded the title of Royal Botanist, a new-fangled term for an emerging science. More to the point perhaps for Parkinson, as a propaganda tool, this large reference work would prove to be second to none.

Paradisi would eventually be recognised as the first English-language book dedicated to horticulture and upon publication its author used it to promote his own view of what a garden could — and should — be to a captive audience who presumably lapped up the advice of the King's botanist.

It is an unashamedly partisan document. Parkinson selected for inclusion his pick of what he considered the very best plants known to man: 'the chiefest for choyce, and fairest for shew, from among all the several tribes and kindreds of natures beauty'.

He recommends around a thousand plants for cultivation — and of those almost one in ten are daffodils.

Parkinson believed he knew the daffodil well. 'There are almost an hundred sorts, as they are seuerally described hereafter,' he writes, describing a variety of European types as well as *N. pseudonarcissus*, which he calls the 'English bastard Daffodill (which growth wilde in many Woods, Groves, and Orchards in England)'.

He wants those who read his book to understand the beauty and potential of daffodils, discussing their scent ('a very few are sufficient to perfume a whole chamber'), and verbally dissecting their form. He elucidates to his readers how diverse they can be, and how some are solitary blossoms while others explode with a profusion of blooms on each stem.

The botanist accuses the 'idle and ignorant' of calling some varieties by the English term 'daffodil' and others by the Latin word 'Narcissus', when these two terms are 'one and same'. He touches upon the problem of how to identify them in the first place, an issue so thorny it would dog botanists for centuries. Ironically, in *Paradisi* Parkinson demonstrates gaps in his own knowledge and misidentifies various other flowers as daffodils.

'There hath beene great confusion among many of our modern Writers of Plants in not distinguishing the manifold varieties of Daffodils,' he writes with anguish, 'if any one shall receive from severall places the Catalogues of their names

... and compare the one Catalogue with the other, he shall scarce have three names in a dozen to agree together.'

In one remarkable passage Parkinson proudly describes how he has grown the 'great double yellow Spanish bastard Daffodill, or Parkinsons Daffodill' in his garden; an account that marks the first known description of anybody ever raising a daffodil not from bulb but from tiny dark seed:

> I thinke none ever had this kinde before my selfe, nor did
> I my selfe ever see it before the yeare 1618, for it is of
> mine own raising, and flowering first in my Garden ... It
> is risen from the seede of the great Spanish single kinde,
> which I sowed in mine owne Garden, and cherished it,
> untill it gave such a flower as is described ...

John Parkinson's love of plants extended well beyond the daffodil and he was, without doubt, an exceptionally able and passionate gardener who swapped all manner of seeds, bulbs and expertise with other high-profile practitioners in England and further afield. Anna Parkinson, the botanist's biographer and a distant descendant, describes *Paradisi* as 'revolutionary' and its author as a Renaissance radical who stood out because he championed plants for more than just their nutritional and curative worth. His vision was that a garden should become 'not just a place of food, sustenance and medical support but a place of beautiful flowers cultivated for their own sake,' she told me, explaining that his true lifework was his encyclopaedic 1640 publication *Theatrum Botanicum* (The Theatre of Plants).

John Parkinson used this book to describe and record the qualities of over 3,000 different plants with such detailed precision it would be long-prized by apothecaries and physicians. Yet by the time he died in 1650 he had fallen from

Gwldr· Grmple·

Grana m· Gvldfir·

Hamp.

Hoppis.

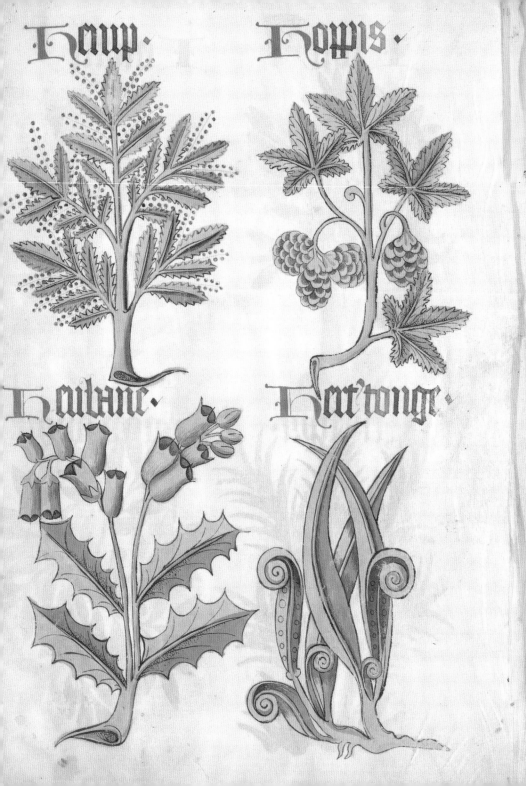

Henbane.

Hartonge.

grace. He would then be sidelined, according to Anna Parkinson who says that when she began researching, relatively little was known about his life.

After his death another odd thing occurred that has long baffled daffodil lovers. *N. pseudonarcissus* continued to grow wild, but virtually every other variety of daffodil John Parkinson so lovingly described seemed to vanish from English gardens, as though they had never been there. It was quite the horticultural mystery.

Anna Parkinson published her findings in her 2007 biography, *Nature's Alchemist: John Parkinson, Herbalist to Charles I*, and argues that this brilliant man's unlikely rise and decisive fall hinged on his being an outsider. John Parkinson grew up in a staunchly Catholic community a long way from London in every respect, his biographer contends — geographically, culturally, economically and politically. Back in 1536 Henry VIII had begun the Dissolution of the Monasteries but some far-flung regions most likely continued with the old, heretical ways, his biographer suggests, deducing that as a boy Parkinson probably learned Latin and Greek from his Catholic teachers. He would likely have seen the effects of the destruction of the monasteries, including the threat that a forest of vital knowledge about plants, and their properties, would be swept away. Parkinson's lifework would, in effect, attempt to right this historical wrong.

John Parkinson's book *Paradisi* can be viewed as a political statement on several levels. It stands as a scientific challenge to the dominant ethos of mysticism, being packed unashamedly with practical, empirical botanical information rather than astrology or superstition, the accepted intellectual currency of the day. Equally significant is its dedication to Parkinson's new Queen Henrietta Maria, and that King Charles I had picked a French-born, Roman Catholic wife.

Anna Parkinson's take on Henrietta, part of the powerful Medici clan, is that she was used to being surrounded by 'wonderful gardens, far superior to anything that was growing in London', which was then by all accounts a rough and ready place. 'The English didn't really begin to get gardens until about the time she arrived,' observes the biographer. 'Obviously there were lilies and things but nothing like the vistas in Europe.'

From the early days of his rule, King Charles fought with his Parliament, which contained a powerful band of Protestant Puritans. In 1629, the year of John Parkinson's *Paradisi*, Charles dismissed Parliament to rule alone. In 1640, as Parkinson published *Theatrum Botanicum*, hostilities between England and Scotland erupted. Charles had to recall Parliament. Despite this, in 1642 England dived into Civil War.

For those whose allegiance lay with Charles these were treacherous times. The writer Robert Herrick, best known for the phrase 'gather ye rosebuds while ye may', was one of the so-called 'Cavalier poets', a band of writers loyal to the King, and he wrote two poems about the daffodil. One, titled 'To Daffodils', interwove church services and the flower's brief life with the need for humans to seize the day. The other, 'Divination by a Daffodil', is a chilling little six-line ditty which at first glance seems simple:

> When a daffodil I see,
> Hanging down his head towards me,
> Guess I may what I must be:
> First, I shall decline my head;
> Secondly, I shall be dead;
> Lastly, safely buried.

In 1649, the year after this poem was published in Herrick's anthology, Oliver Cromwell signed Charles's death warrant and had him publicly beheaded at Whitehall in London. A superstition circulated that daffodils were cursed, and that those who saw a daffodil droop were likely to die.

Catholics were ordered to leave the city but somehow John Parkinson managed to stay on, says his biographer, and in so doing must have seen everything he held dear crumble around him. The king's botanist died the year after his king with no one left who would have dared celebrate the achievements of his remarkable life.

There was 'no state ... no monarchy, no army', Anna Parkinson told me. 'He died in this state of complete anarchy.' John Parkinson's achievements appear to have been overlooked to a degree by historians after his demise. His biographer argues that this happened because he was Catholic, and therefore intrinsically 'at odds with the politics of his day'.

Is it possible that the daffodil became swept up in this, I wonder. Many varieties came from Catholic countries such as Spain, France and Portugal, so to the English the flower may have became a dangerous emblem, symbolic of political dissent. Did the daffodil become a silent victim of England's bloody civil war?

Anna Parkinson agrees that everything that took place at this time was connected with the politics of the day but suspects that the daffodil's disappearance had more to do with fate than conspiracy. She makes a good point, though given the daffodil's association with Catholicism, I am not so sure.

After Parkinson's death almost all of his daffodils disappeared from the gardens of England, Wales and Ireland. *Narcissus pseudonarcissus* survived and would be 'officially' named (it translates, rather bizarrely, as 'daffodil false-daffodil') in 1753, when Swedish botanist Carl Linnaeus published the first systematic taxonomy of living things in his masterwork *Species Plantarum*, the pages of which also name the genus *Narcissus*.

In the 1800s daffodil hunters began to find, here and there, a few examples of ancient varieties growing near the sites of monasteries and other religious buildings in England and Ireland. They surmised that monks had imported these flowers from Catholic countries such as Spain and Portugal, used them for herbal medicine and that a few particularly hardy strains had survived the devastating Dissolution of the Monasteries.

Across Britain, in hedgerows, woodlands and lakesides, feral *Narcissus pseudonarcissus* continued to grow wild — an outsider, perhaps even a rebel, and it would take considerable effort to change that. In 1813 Lord Edward Hovell Thurlow, a contemporary of William Wordsworth, published a verse about the 'pale narcissus' in his book *Poems on Several Occasions,* in which he described the daffodil with utter disdain, as a flower that fed upon itself. He wrote that the nymphs of old would pluck such daffodils 'from its tender stalk. And say, "Go, fool, and to thine image talk".'

Since then *Narcissus pseudonarcissus* has become ingrained in modern life — quite literally in 1999 when European researchers genetically spliced some of its DNA into the prototype of Golden Rice, a controversial cereal aimed at combating Third World malnutrition.

In its wild state, in the untended sections of my mother's garden, the daffodil is a thing of beauty that tells her heart spring is here. Her notes describe *Narcissus pseudonarcissus* as a deceptively modest flower: 'a small Div I W/Y trumpet daffodil, that with "Poeticus" bred all the daff cultivars. Pale yellow or white petals with strong, yellow trumpet.' She calls it 'Wordsworth's daffodil' still.

CHAPTER 2

Daffodilians — A Potted History

'Daffodil growing is a cult. To grow daffodils once is to grow them always.'

'The Spring Show',
New Zealand Herald, 4 September 1908

It is impossible to understand what daffodil lovers call the 'modern' daffodil without meeting some of the nineteenth-century characters who created it and who, in so doing, re-engineered spring. William Herbert (1778–1847) is the first and most intellectually intriguing; a politician, linguist, poet, clergyman and cerebral firebrand who came to be known by a nickname he adopted for himself, which to the modern ear has a mafia-esque ring to it — 'the Dean'.

Herbert belonged to England's powerful, wealthy elite. At his birth his politician father Henry Herbert (1741–1811), who would be given the title Earl of Carnarvon by King George III, had possession of Highclere Castle in Berkshire, his clan's family seat. Henry Herbert had taste and ambition. He commissioned

Capability Brown (1716–1783), England's most revered landscape architect, to design the grounds of Highclere, where young William would run and play.

Highclere imprinted itself on the public's imagination in 1922 when the fifth Earl of Carnarvon and his archaeological colleague Howard Carter discovered the tomb of Egypt's boy king Tutankhamun, an apparently curse-laden adventure that inspired countless wide-eyed newspaper articles, books, movies and conspiracy theories. In 2010 Highclere Castle hit the world stage again as the filming location for *Downton Abbey*, a British television period drama about the trials and tribulations of a fictitious Edwardian family and its loyal staff. *Downton Abbey*'s success helped save the crumbling Highclere, but two centuries earlier it was William Herbert's future that seemed unclear. As child number five and son number three the young aristocrat had scant prospect of inheriting, which left a smattering of socially acceptable career options open to him — politics, religion, the army and the law.

Schooled at the elite Eton College, Herbert attended Oxford University's Christchurch and Merton colleges and chose the path of power, although two brief stints as a Member of Parliament (first for the constituency of Hampshire then for Cricklade) prompted a change of tack. He decided that the Anglican Church might suit him better and took holy orders in 1814. At the age of thirty-six, thanks to an obligingly nepotistic family member, he became rector of a sleepy Yorkshire village called Spoffoth.

The position seemed perfect. It guaranteed Herbert an annual salary of £1,600 (no small prize given that the average labourer's yearly wage sat at around £40), a comfortable vicarage home with gardens, and plenty of time, which he used to tackle one of the most exhilarating and potentially heretical challenges of his day: the riddle of life itself.

By 1840 Herbert had taken up the prestigious post of Dean of Manchester (from whence came his nickname) and undertaken decades of systematic

experiments on plants. He collated data, published research, battled with creationists and impressed his elite peers, a group that included a brilliant youngster called Charles Darwin. In the process Herbert became Britain's first known amateur plant hybridiser and of all the species he worked with none was closer to his heart than the daffodil.

The Dean's world was the first true era of information overload. Explorer-collectors roamed the globe, returning home with an apparently unquenchable flood of discoveries from the natural realm; remarkable finds that ranged from exotically sensual fruit to exquisite sea animals and the relics of bizarre, monstrous beasts.

This onslaught of extraordinary data had far-reaching cultural and economic consequences. A lucrative market emerged; indeed some London auction houses dealt with nothing but highly sought-after imported exotics, and the blitz of discoveries presented a provocative challenge for the curious who understood that collecting examples of this massive diversity of life was one thing but bringing systematic order to the apparent chaos they represented was quite something else.

Could it possibly be, some brave souls speculated, that a hidden mechanical process linked beasts of the ancient past to those still living? What did apparently unrelated animals and plants have to do with each other, and where did God fit into this? If living things had somehow evolved over time, just how much time had it taken — and what was the implication for human beings?

Two types of men stepped up to the challenge: professionals and wealthy amateurs, a fair number of whom, like William Herbert, were helpfully employed by the church. It is no coincidence that the word 'scientist' was minted in this heady era by an Anglican priest called William Whewell, a distinguished Cambridge University philosopher and historian who wanted to replace the confusingly amorphous phrase 'natural philosopher' with something less cumbersome and more precise.

Herbert, like Whewell, possessed the dexterous mental agility of a polymath. He published poetry he had translated from Greek, Latin, Danish, German and Icelandic to English, as well as writing lengthy volumes of his own epic verse. His true life work, however, emerged from the results of his systematic examination of the sex lives of living things. Herbert used his vicarage gardens as a botanical laboratory where he could forensically experiment with vegetables and ornamental flowers, and attempt to pioneer the science of botanical cross-breeding.

Farmers had long understood that plants and animals could be deliberately crossed to produce useful traits, but the mechanics behind this remained mysterious at best and the notion of it, to many people, profoundly frightening. The issue of how to classify living things properly was another area of bewilderment, as was whether — as most believed — creatures produced by mating different kinds of animals or plants (such as asses, born when horses and donkeys cross-breed) were always infertile.

From as early as 1819 Herbert's written work made it clear that he knew better. His first essay for the Royal Horticultural Society (titled 'On the Production of Hybrid Vegetables') outlines his ideas, challenges the accepted plant classification systems of the day and describes some of his techniques, which include making diligent microscopic examinations of botanical cross-sections. He believed, quite rightly, that hybridisation held enormous potential and could be transformational for humankind.

'The vegetable kingdom may certainly be greatly enriched by artificial inter-mixtures,' he deduced. 'I think that new plants may be so formed, which will be capable of reproducing themselves.'

Herbert experimented with many plants but was particularly intrigued by *Narcissus* and perplexed by the fate of John Parkinson's mysteriously absent daffodils. He wondered whether wild French daffodils were natural hybrids, and with *N. pseudonarcissus* relatively easy to find in the English countryside he sourced a handful of other ancient varieties, such as *N. poeticus* and *N. incomparabilis*, and started forcing them to breed with each other.

He describes his painstaking technique thus:

> The six anthers should be carefully taken out before the flower which is to bear the seed, blows. This may be done through a slit cut in the tube; and the yellow dust from another sort must be applied to the point of the style.

Herbert discovered that creating new daffodils was boundlessly exciting. He became entranced by the unexpected new forms and vivid colours his crosses produced and urged others to try it, calling 'the attention', as he put it, of anybody 'who has a spot of garden, or a ledge at his window, to the infinite variety of Narcissi that may be thus raised, and most easily in pots at his window, if not exposed too much to sun and wind, offering him a source of harmless and interesting amusement, and perhaps a little profit and celebrity'. The Dean's passion for hybridisation went far beyond any single plant. He wanted to understand the complex connective web that he believed linked creatures of the past to those of the present; the mechanisms by which organisms changed through generations and over time. In 1837 he published *Amaryllidaceae: Preceded by an Attempt to Arrange the Monocotyledonous Orders*, a dense work (as much of his is) packed with findings, conclusions, speculations and the unexpected evidence that, on top of everything else, he possessed unusually accomplished drawing skills. His images are quite lovely.

Some botanists feared that intermixing species would create a wilderness of uncertainty. Herbert argued the reverse; that engaging in careful cross-breeding

would force the cultivator 'to study the truth, and take care that his arrangement and subdivisions are conformable to the secret laws of nature'.

He rightly viewed the plant world as part of a much larger evolutionary adventure and his speculations touch topics as diverse as the behaviour of birds, the emergence of languages around the world, and what he controversially termed 'the origin of the different races of mankind'.

Herbert had no doubt that at the very least mammals were intrinsically, biologically connected, a prescient view considered heresy by many, not least in his religious world. 'I feel satisfied that the fox and the dog are of one origin,' he wrote, touching on the idea of 'reversion', that creatures could demonstrate throw-back characteristics of a distant, ancestral past:

> Nor could I ever contemplate the black line down
> the back of a dun pony without entertaining a suspicion
> that the horse, unknown in the wild except where
> it has escaped from domesticity, may be a magnificent
> improvement of the wild ass in the earliest age of
> the world.

Charles Darwin (1809–1882), a twenty-eight-year-old botanist, feasted on Herbert's book. The previous October, Darwin had returned from his confounding five-year global exploration aboard HMS *Beagle*. He brought back botanical specimens, wonder-laden trip notes, and an overwhelming urge to understand what had generated the enormous variety of natural life he had been witness to; the 'mystery of mysteries', as he described it.

Darwin interrogated many eminent researchers about their work and ideas. He annotated his copies of Herbert's publications, kept them close and bombarded the Dean with questions, in letters and in person, visiting at least twice (in 1845 and 1847).

The issues that transfixed Herbert equally fascinated Darwin, and some of the letters between them are considered among the most significant of Darwin's correspondence on natural selection. A third man figures in their associations: John Stevens Henslow, a cleric and professor of botany at Cambridge University, whose actions altered the path of Darwin's life.

Henslow appears to have introduced Darwin to science (the younger man had originally gone to Cambridge to study theology and the classics), and Henslow secured Darwin's place on the *Beagle*. Together, the three men then effectively defined the boundaries of the work Darwin would dedicate most of the rest of his life to.

Darwin clearly found the Dean's methods and theories stimulating. The older botanist argued that plant species that can no longer be forced to intermix have 'descended from one original' and he suspected 'such a multiplication of distinct species' had also taken place in 'the animal and insect tribes'.

Such heretical notions made the cleric deeply unpopular with some of his brethren. The footnotes in *The Christian*, the Dean's 1846, final volume of poetry, sizzle with frustration at the idiocy — as he beheld it — being voiced by those he derided as 'sticklers for the literal meaning of words in the Scriptures':

> He [God] created the material globe, vegetable, fish,
> reptiles, insects, birds, quadrupeds and man, but ... He
> may have impressed upon them a principle of change
> and development ... Everything that has come within my
> knowledge leads me to think, that, although the earth
> and its productions are now in a very quiescent state, a
> powerful principle of change was in operation in the first
> ages after the deluge, when the animal and vegetable
> races were dispersed over the varied surface of the globe.

It would be preposterous to suppose, that the words [of
the Bible] were meant to be understood literally.

Herbert died in London on 24 May 1847. Twelve years later, on 24 November
1859, Darwin published *On the Origin of Species by Means of Natural Selection*, his
revolutionary book on evolution. Within its pages Darwin pays tribute to the
Dean's hybridising expertise and 'great horticultural skill', discusses at length
the older scientist's experimental findings, and on the topic of 'the struggle for
existence' states that, 'In regard to plants, no one has treated this subject with more
spirit and ability than W. Herbert, Dean of Manchester, evidently the result of his
great horticultural knowledge.'

Decades after the Dean's death, daffodil lovers lobbied for a reassessment
of William Herbert's life and work. Irish explorer and plant collector Frederick
William Burbidge (1847–1905), in his 1875 book *The Narcissus: Its History and Culture*,
describes the Dean as being half a century ahead of his time in his understanding
of 'hereditary descent'. 'And this,' notes Burbidge with clear-eyed admiration,
'when many botanists strongly and openly objected to the practice of hybridisation,
and years and years before the publications of Charles Darwin.'

In 1899 the 18 March edition of *The Gardeners' Chronicle*, the influential British
illustrated gardening journal, hailed the Dean a 'versatile genius ... a pre-Darwinian
Darwin', and ninety years after his death the English geneticist Cyril Dean
Darlington described his contribution to evolutionary science as 'the thin end of
the wedge which Darwin drove home'.

The Dean was right — hybridising would revolutionise farming as well as
gardening, and for those captivated by the daffodil he had prized open the door to
an exhilarating new world.

*There is no end to the varieties and elegant forms
that may be obtained.*

Edward Leeds, quoted in *The Latest Hobby: How to Raise Daffodils From Seeds*, 1908

Edward Leeds (1802–1877) burst onto the daffodil scene in 1851 with a call to arms in an essay for Britain's *Gardener's Magazine of Botany* in which he announced: 'There is no end to the varieties and endless forms that may be obtained.'

Leeds knew this from exhilarating experience. Years of cross-breeding had produced breathtaking results in the form of six 'new and beautiful varieties' of *Narcissus* — *N. Leedsii* (named after himself in what would become a hybridisers' tradition), *N. poculiformis elegans, N. major superbus, N. aureo-tinctus, N. incomparabilis expansus* and *N. bicolor maximus*, described as the most symmetrical and perfect of all the daffodils.

'I think much remains to be done in the production of fine hybrids of this beautiful tribe of plants,' Leeds declared. 'These are not ephemeral productions, like many modern florists' flowers, but will last for centuries with very little care.'

Leeds packed his essay with practical tips explaining: how to coax late-flowering varieties into opening before their time so their pollen can be used with early-flowering daffodils; how to source outstanding soil; and how to care for one's charges through the winter until their bulbs have safely matured. He makes clear that daffodil breeding takes time and perseverance.

When the bulbs are two years old, prepare, in an open
airy situation, a bed of good loam, mixed with sharp sand,
in which the bulbs should be imbedded; plant the roots
in rows, 3 inches apart. They will stand three years in
this bed. A few will flower the fifth year, but the greater
portion not until the seventh.

Slowly, carefully and very methodically Leeds hybridised around 150 different daffodil varieties that included *trumpets, tazettas, incomparabilis, biflora* and *poeticus*. He cultivated his offspring in pots under glass, reputedly never let anybody else as much as touch them, and scoured Europe from his living room via any plant list or catalogue he could find.

He sought out enthusiasts to correspond with, combining the seditious single-mindedness of a scientist with an aversion to travelling that appears to have bordered on the agoraphobic. Even so, he made himself known to some of the most eminent botanists in England and assembled an extensive spectrum of plants, the most outstanding segment of which was his unique, self-generated daffodil collection — hundreds of new varieties and in excess of 20,000 bulbs.

So how did this all begin?

Much of what is known of Leeds stems from a profile penned by horticultural enthusiast William Brockbank for two mid-November 1894 issues of *The Gardeners' Chronicle*, the plant-lovers' bible. Brockbank never met Leeds but made an early spring pilgrimage to his home in Longford Bridge, Stretford (about 6 kilometres from Manchester), shortly after his death. Leeds's garden appeared 'ablaze with Daffodils — growing by the thousands, almost wild', reported Brockbank, who grilled local gardeners for information to build up a picture of a rather curious man.

Born on 9 September 1802, Edward Leeds was oldest of four in a family that lived in Pendleton, near the enormous polluted factory town of Manchester. Dubbed 'Cottonopolis', this nineteenth-century slum-laden, smoke-stacked powerhouse came to represent Britain's Industrial Revolution at its most brutal.

Edward Leeds's father, Thomas, was a cotton mill owner, and the boy's reaction to his environment appears to have emerged early, judging from his schoolyard nickname ('Bulbous'), and his inclination to spend his pocket money on plants.

In February 1829, Thomas hurtled into bankruptcy. Edward, already winning prizes for his herbaceous plants at local flower shows, agreed to join forces with his father in a new share-broking firm, which appears in the Manchester Directory as 'Leeds, Thomas & Son' three years later. Edward married, moved to nearby Stretford, where he would spend the rest of his life, and fathered the first of four boys, three of whom would survive to adulthood and all of whom would qualify as doctors.

Thomas's death in 1839 left Edward Leeds in full control of the family firm. By now he was writing monographs on a range of plants to broadcast information and practical tips gleaned from his own experience. He collected dried flora, wrote letters to, and exchanged plant specimens with, some of the top horticulturalists of the day, and cultivated everything from amaryllis to peonies. His garden became a magnet for elite breeders, and on that first visit Brockbank seems overwhelmed with delight at what he saw:

> At one place the Veronicas had prevailed, at another the Saxifrages had gained the ascendancy. There were large clumps of the most beautiful *Orobus vernus* I had ever seen. Hardy Geraniums had grown into huge masses … Of the smaller Irises, there were large clumps. Creeping plants had trailed over the walks, and in the borders were thickets of huge growers. Telekias, Asters, Campanulas, Delphiniums, Veratrums, Heracleums, and other giants, in rank profusion, the fittest only having survived.'

Brockbank said he spent months carefully annotating exactly what was growing then uprooted the best specimens, making sure he included Leeds's unique daffodils. The writer took them to his own garden and planted them. There, he noted with satisfaction, 'they now abound'.

A gardening friend of Leeds's told Brockbank how the hybridiser would stand guard over his daffodils waiting for the moment the flowers emerged. Then he

would remove the stamens, fastidiously apply the pollen of his choice as soon as a pistil ripened, and seal up the fertilised daffodil flower by entwining it with thread to ensure no pollen-laden bee could interfere. Leeds's seedling beds contained hundreds of daffodil varieties, and he 'never tired of descanting on their beauties'.

Few shared this man's deep devotion to the flower and that left Leeds infuriated. By the early 1870s his health was deteriorating. Unable to walk, he found himself reduced to visiting his garden by being pushed there in a three-wheeled chair. He spent sunny days watching over his creations, and considering how best to protect his life's work.

Leeds offered to donate his herbarium to Kew Gardens, and the institute accepted, but barely anybody seemed interested in taking on his daffodils. He let it be known that his wife was under orders to destroy his entire *Narcissus* collection as soon as he died — a threat designed to flush out somebody prepared to 'adopt' his daffodils. More on that later. Leeds's death coincided with daffodil season and the height of spring on 4 April 1877. To Brockbank the knowledge that Leeds did not live long enough to see many of the daffodils he had created bloom, seemed poignant indeed.

Aim high, and do not waste your time in producing inferior flowers. It is the same trouble to raise a useless variety as it is to raise a high-class flower.

Peter Barr, admirer of William Backhouse, in a speech to the Sea Point Horticultural Society, Cape Town, South Africa, September 1901

William Backhouse (1807–1869) was a Quaker from County Durham in the north-east of England. There is nothing to indicate he ever met fellow pioneering enthusiast Edward Leeds yet their stories contain echoes of each other: both born around the same time to northern English families, both destined to die before their achievements would be fully understood, and both doomed to have their life work threatened.

Backhouse carved out his career in the offices of Backhouse's Bank, his family's business in Durham. Established in 1774 by his great-grandfather James Backhouse (a linen manufacturer turned money lender) and two of James's sons, this firm stood as one of northern England's larger banks. Like many of his kin William Backhouse had a keen fascination for the natural world, and by the age of twenty-two he had become a founding member of the Natural History Society of Northumberland, Durham and Newcastle-upon-Tyne, today one of the oldest English organisations of its kind with the less cumbersome name of 'Natural History Society of Northumbria'.

Backhouse would turn his investigative attentions to insects, birds, geology, and meteorology and upon experimenting with his first daffodil cross in 1856, at the gardens of his home, St John's Hall, near Wolsingham, became utterly entranced.

The mechanics of heredity fascinated him. He observed with no little puzzlement that first generation daffodil 'children' often look nothing like either of their parents, and set to producing flowers of different shapes, sizes and colours that ranged from pure whites, lemons and yellows to some with red edges and glowing orange tints.

He learned fast and experimented extensively, generating dozens of new varieties with a clear idea of what he wanted to achieve. A man of routine, the banker hybridised first thing in the spring mornings before catching his train in to work, placing his flowering daffodils in pots in a miniature glass structure, more porch than greenhouse, connected to his library.

The results were thrilling. In 1865 Backhouse sent a box of seedlings to *The*

Gardeners' Chronicle — which the editor judged to be 'of remarkable beauty' — with a reminder that 'Narcissus are among the most varied and interesting of early-flowering spring bulbs', an indication that the horticultural cognoscenti were beginning to see the potential of this flower.

The *Chronicle* devoted space in its 10 June issue to Backhouse's experimental methods and remarkable crop. 'The Daffodils Narcissus major, Ajax, Pseudo-Narcissus, minor, and moschatus, cross with one another, and they produce seeds as freely as the parents,' Backhouse explained. 'The colours are not merely intermediate, but of all shades between the colours of the parents.'

In the spirit of encouraging new breeders, he outlined his technique which mirrors that of Leeds. 'Before dusting with pollen I cut out the stamens, except in the kinds which only seed when crossed' Backhouse wrote, '...but in out of door plants sometimes the bees may be the authors of a different cross. I have sometimes tied the mouth of the cup in the Daffodils for a time to prevent access.'

By crossing two wild species — *N. bicolor* (which he used as the seed parent) and *N. pseudonarcissus* (his pollen parent) — Backhouse produced a pair of breathtaking daffodils he named 'Emperor' and 'Empress'. These arresting, long-cupped trumpets are now categorised as Division 1 daffodils and Daffseek. org, the American Daffodil Society's handy digital daffodil database, describes 'Emperor' thus:

Perianth [petal] segments ... brilliant greenish yellow, with a darker tone at midrib beneath, somewhat inflexed, plane, overlapping one-third; the inner segments with margins wavy; corona cylindrical at base, flared in upper part, loosely ribbed, vivid yellow, with mouth expanded and lightly frilled, rim flanged, notched and crenate.

Backhouse could not have known it but he possessed something of a Midas touch when it came to breeding 'triploid' daffodils, that is Narcissi possessing twenty-one chromosomes (three sets of chromosomes) rather than the more typical fourteen (two sets of chromosomes, known as 'diploid'). 'Emperor' and 'Empress' are two of the first such daffodils ever cultivated, and their impact would be far-reaching but the banker went further still, creating a magnificent, tall, white-yellow bloom called 'Weardale Perfection', the first known 'tetraploid' daffodil with four sets of chromosomes. That level of genetic diversity would result in tetraploids dominating the 21st-century daffodil, something Backhouse could never have foreseen.

Thirteen years after making his first daffodil cross Backhouse died. At first everything he had so carefully amassed — his specially bred bulbs, accumulated seeds, and cross-breeding notes — seemed to languish at his home.

In time he would be recognised, as one of the 'celebrities of Daffodildom' as the 1913 *Daffodil Yearbook* put it, yet his legacy did not end there.

Backhouse began the first true daffodil dynasty. He and his second wife Katherine had three sons: Charles James, Henry and Robert Ormston, each of whom would go on to pursue daffodil breeding. Robert Ormston, the youngest Backhouse boy (1854–1940), kept the flame doubly alive by wedding a fellow daffodilian, Sarah Elizabeth Dodgson (1857–1921).

As Mrs R.O. Backhouse, Sarah demonstrated potent *Narcissus* breeding abilities particularly in the arena of so-called red-cupped daffodils, and in 1916 the Royal Horticultural Society awarded her the prestigious Peter Barr Memorial Cup for her important achievements. Two years after her death in 1921 her widower Robert christened 'Mrs R.O. Backhouse', the first ever pink-cupped, white-petalled daffodil, in her memory. A snowdrop, 'Mrs Backhouse's Spectacles', was also named for her.

Guy L. Wilson (1885–1962), an eccentric Irish narcissophile, paid tribute to Sarah's talents by recalling one of her 'sensational' flower exhibits at a particular show, 'for the most part arrayed in flaming colours, in one or two instances quite barbaric in effect. One could rarely see the flowers, such a crowd of admirers besieged them all the time,' he observed, and was quoted in American gardening authority Mrs Francis King's 1925 volume *Chronicles of the Garden*:

> I think Mrs. Backhouse must have a feeling for dramatic effect, and of keen appreciation of the value of climax from the way in which she unpacked these flowers. She kept quietly putting up one wonder after another, amid a crescendo of superlatives from the onlookers; thinking she had arranged all her flowers, I left her stand, but passing it again a little later, I saw in the centre three flowers which reduced me to incoherent amazement.

'Empress' was one of the heirloom daffodils identified by the bulb hunter who visited my mother's garden. It must have been growing there since at least the 1930s yet it blooms with a youngster's nimble pep, all tender white petals and glowing yellow trumpet. My mother decided it could do with some company, and sought out six bulbs of 'Mrs R.O. Backhouse' and another three bulbs of 'Emperor' to plant alongside.

The admirable Butterfly

CHAPTER 3

The Daffodil King

'Peter Barr, in no figurative sense,
made the Daffodil ... The "King" as
we like to call him, lived long.'

Reverend Joseph Jacob,
Daffodils ... With Eight Coloured Plates, 1910

'Barri Conspicuus' is a handsome little devil. It sports paper-thin yellow petals
that fade in colour as they age, shimmering golden cups laced with a decidedly
grandiloquent scarlet, and has grown in my mother's garden since she spotted a
photograph of it in Ron Scamp's catalogue and sent away by post for ten bulbs.
She planted them in a sun-dappled woodland slope behind a flowerbed overrun
by grapevine, made a note of the spot and jotted down what she knew about
the daffodil: '3Y/Y YO (LM) Pre 1921.10 for £5.95.' 'Barri Conspicuus' has since
demonstrated a headstrong character and run rather wild.

   This flower is named for arguably the most significant cast member of the
Narcissus drama: a stocky Scotsman called Peter Barr (1826–1909) who was destined
to become known as the 'Daffodil King'.

PETER BARR, V.M.H.

'The fact is that the world would not have had the daffodil but for me,' Barr bluntly declared to the *Bendigo Advertiser* for a story, published 23 November 1900, on the antipodean leg of his self-styled daffodil world tour. As Barr told it, his passage to floral royalty was practically preordained. 'I was born within a few yards of a tulip bed,' he revealed. 'I have been amongst flowers ever since.'

Those tulips had blossomed in Gowan, a Scottish parish destined to become part of the nearby city of Glasgow, and were the handiwork of Barr's father — a mill owner who cultivated them in the family garden and whose talent for gardening appears matched only by his capacity for producing sons.

Peter, born 20 April 1826, was the seventh of twelve Barr boys. By his tenth birthday the family textile business had foundered, forcing the youngster to leave school and help out financially. He worked briefly with a weaver before beginning as a junior at a city seed store.

Barr would judiciously prune his personal history as time went by, once suggesting that it was then, as a stripling, his relationship with daffodils truly began. What is clear is that horticulture thoroughly suited him. He worked his way up to manager, moved to Ireland to work with another seed-selling firm and, upon detecting the sweet scent of opportunity, transplanted himself to the English city of Worcester.

A photograph of Peter Barr taken at the age of twenty-two reveals a dapper and determined young man, attired in starkly white shirt with upstanding collar and a large, softly fashionable bow tie. Four years later Barr launched his first horticultural firm with the help of a business partner. The venture lasted just four years and coincided with his meeting an Englishwoman called Martha Hewlings. By the end of the 1850s they had wed (a union that delivered at least seven children) and relocated to London, a city that boasted two million denizens and the distinction of being the largest metropolis on Earth. Barr wanted to work at the epicentre of Britain's flower industry.

The Scot moved fast. He launched Barr & Sugden with fellow seedsman Edward Sugden at 12 King Street, Covent Garden. That same year, 1859, British journalist George Augustus Sala wrote of the vibrant Covent Garden market in a special investigation called 'Twice Round the Clock, or The Hours of the Day and Night in London'.

Sala saw buyers wanting to cherrypick the finest fruit and vegetables streaming into the market from 6 o'clock in the morning. 'Sweeter even than the smell of the peas, and more delightful than the odour of the strawberries, is the delicious perfume of the innumerable flowers which crowd the north-western angle of the market,' he wrote, 'from the corner of King Street to the entrance of the grand avenue.'

An ocean of petals confronted Sala: 'hundreds upon hundreds of flower-pots, blooming with roses and geraniums, with pinks and lilacs, with heartsease and fuschias ... mignonette and Jessamine ... peculiar roses with strange names ... [and] cut flowers, too, in every variety'.

In such a jungle of horticultural capitalism only the fittest entrepreneurs could survive, and Barr was taking no chances. Within two years his company broke new ground by publishing catalogues displaying not just the expected verbal descriptions of flowers on offer but daring, wood-engraved illustrations as well. His audacity attracted instant criticism but would be seen as pioneering; illustrated horticultural catalogues became, and remain, standard practice worldwide.

Barr wanted to do more than simply sell other people's flowering bulbs. He established a nursery in Surrey that doubled as his laboratory and studied ornamental garden plants, including the origins of lily, tulip and the evergreen hellebore cultivars. He drafted letters to those he hoped might assist, a list that grew to embrace celebrated breeders, professional botanists and seedsmen from Britain to the Netherlands.

Decades later Barr regaled audiences with fond memories of that exhilarating, stressful time, depicting it as an era when daffodils attracted little more than loathing. People might condescend to part with a tiny sum of money for a handful of cut wild daffodils every now and again from a roadside seller, but the British public despised *Narcissus* as nothing more than a flowering weed. It was a 'tabooed flower' he said, one that 'few people would tolerate ... in their garden'. The Scot noted this, thought it terrible, decided it represented the entrepreneurial opportunity of his life, and that tastes must change.

A more dispassionate view of events can be gleaned by studying the fascinating, densely packed pages of *The Gardeners' Chronicle*. Founded in 1841 it covered every aspect of horticultural and gardening news, at home and abroad, with earnest, often droll contributions from scientists, luminaries, amateurs and hobbyists alike.

In the first issue of 1869 only one company took out a paid advertisement in the *Chronicle* to tell readers it had imported 'Polyanthus Narcissus' for sale.

'Imported from where?' a reader might wonder. That question would be answered in subsequent issues, where paid notices alerted canny retailers about upcoming auctions featuring cases filled with hyacinths, tulips, crocuses, jonquils and narcissus bulbs fresh off ships from Holland.

Daffodils clearly captured the attentions of *Chronicle*'s editorial staff who used its 3 April Spring issue to pose a very basic question: 'What is a species of *Narcissus*?' An impetuous proposal to mount a daffodil exhibition had thrown up all manner of unknowns from what these strange blooms were to who might be qualified to judge them in a competition. Nobody, it seemed, knew enough to do it. As the publication put it: ' Seriously, we don't suppose any three botanists would be of the same opinion as to what constitutes a species in this more than usually variable genus.'

The *Chronicle* decided to take the matter in hand. It announced that John Gilbert Baker (1834–1920), botanist at the Herbarium, the Royal Botanic Gardens, Kew, had agreed to tackle these critical questions and urged readers who might have 'specimens in a fit state for determination' to send them either to the *Chronicle* or directly to the botanist. Baker then attempted to review, describe and collate a *Narcissus* classification that could include every variety being grown in Britain.

His first set of results, outlined in a *Chronicle* essay published on 17 April 1869, noted 'the reviving interest in this beautiful genus' and paid tribute to two men: the late botany-loving scholar Adrian Hardy (A.H.) Haworth (1767–1833), who had printed a monograph two years before his death which identified 150 species under sixteen genera, and William Herbert (the Dean), who had taken it upon himself to shrink Haworth's species list to six.

Baker argued that the tally of 'what may be fairly called species … which are known definitely in a wild state' was in fact 'much over twenty' and after carefully assessing their outward appearance proffered an innovative classification framework designed to cut through centuries of confusion.

Having categorised each flower according to the relative lengths of its two main visible features — its perianth (defined as the sepals, or calyx, and petals, or corolla) and its crown (variously also called the corona or trumpet) — Baker divided up the *Narcissus* varieties he had studied into three groups: 'Magnicoranti' (flowers with crowns as long, or longer, than their perianth), 'Mediocoronati' (crowns half the perianth length) and 'Parvicoronati' (crowns less than half the length of the perianth). Baker had no idea that his smart new classification system would last merely decades.

It is clear from the pages of *The Gardeners' Chronicle* that Britain faced an existential crisis. Correspondent Alexander Forsyth published an article in the periodical's 8 May issue depicting the extent of inequality in 1869 England, and the deep scars of industrialisation. Beyond London many of the poor lived in districts smothered in 'fire and smoke (not chemical works' effluvia, but honest burnt clay smoke)', Forsyth wrote, noting that working class life in the capital city seemed barely any better 'where poverty is an ingredient … as well as smoke'.

Hawkers trying to sell flowers had long traipsed their donkeys around England but there seemed to be a change in the air. The days when growers scorned these humble traders were over. 'Thousands of plants are grown for hawkers now,' Forsyth observed, identifying a hierarchy to blossom buying: 'The Hyacinth, the Crocus, and the Narcissus belong to the better class of buyers and do not come within the range of the poor man's purse, for the bulbs being worth 6*d.* [sixpence] to 9*d.*, and the pot 1*d.* or 2*d.* more, the whole thing is a luxury.'

The daffodil appeared to be attracting some powerful allies. Miles Joseph Berkeley (1803–1889), a Cambridge-educated vicar, exotic fungal researcher and pioneer in plant pathology, directed considerable energy lobbying for it. He urged the Royal Horticultural Society to launch a daffodil show and let it be known that Lady Dorothy Nevill, a writer, horticulturalist and darling of the social scene, would donate a special prize for the best *Narcissus,* as reported in *The Gardeners' Chronicle*, 22 May 1869. Nevill had influence and no qualms about using it. Her contact book included everyone from Disraeli to Kew Garden's Joseph Hooker and even Russia's Catherine the Great, and her passion for the natural world and for science seemed inexhaustible. She collected exotic birds, farmed silkworms, owned conservatories bristling with rare plants and attracted the attention of scientists such as Charles Darwin, to whom she sent orchids in the hope of furthering his research. When it came to the subject of supporting daffodils — such a sweetly alluring alien breed — she may not have needed much persuading.

In letters to her friend Lady Airlie, Nevill divulges her excitement about finally meeting Darwin ('I am sure he will find I am the missing link between man and apes') and how *Narcissus* could lift her spirits. 'It is bitter winter; but I have been surrounded by primroses, daffodils, and violets — a land of promise truly,' she wrote. 'This later on will indeed be a land of flowers. I have such lovely pansies and anemones and every sort of daffodil.'

Change was in the air in the form of a grassroots revolution of the most fundamental kind. In 1870 William Robinson (1838–1935), an Irish horticulturalist who would become recognised as the father of the English flower garden, generated the spark with his publication of *The Wild Garden*, a gardening book that voiced a battle cry. Within its pages Robinson encouraged people to re-evaluate the world around them by rejecting the straitjacket of formal gardens and shunning what he called 'repulsively gaudy' city park flower planting. Instead, he wanted his readers to embrace the quasi-chaotic wilderness of Britain's untended, natural landscapes.

'We go to Kew or the Crystal Palace to see what looks best there ... or the weekly gardening papers tell us; and the following season sees tens of thousands of the same arrangements and patterns scattered all over the country,' wrote Robinson, arguing that true beauty lay in the 'infinite delight' of the 'semi-wild'.

Linking these 'paradises of vernal beauty' to the great dramaturge William Shakespeare and the land's most compelling poets, Robinson proposed a radical reassessment of Britain's native flowers, recommending that gardeners mix 'the best hardy exotics' with 'the best of our own wild flowers in wild or half-wild spots near our houses and gardens'. The outcome would be 'the most charming results ever seen in such places,' he wrote, adding, 'To most people a pretty plant in the wild state is more attractive than any garden denizen. It is free ...'

In a culturally holistic sense it is difficult to separate this call to arms from the ground-level trauma of the Industrial Revolution, which had wreaked such havoc on Britain's landscape, its working class, and its psyche. Robinson's precocious book predated the Arts and Crafts movement by about a decade. It foreshadowed a fashionable passion for creating idyllic English country gardens — a phenomenon that now appears timeless — and as Robinson identified flowers of particular interest, be they lilies, bluebells, foxgloves or violets, he made a point of singling out the daffodil.

In listing eleven members of the 'Daffodil Family', some incorrectly, Robinson noted that all appeared to be foreigners, hailing variously from Southern Europe (the Hoop-Petticoat Daffodil, Sweet-scented Daffodil and Poet's Daffodil), Spain (Small Daffodil, Great Daffodil) and Portugal (Peerless Daffodil). Even the familiar, fragrantly bunch-flowered, wild-growing 'Poet's Narcissus' (*Narcissus poeticus*) is 'not considered truly British' despite the profusion of pedlars that bloomed each spring to sell it on the streets of London, Robinson wrote. He held *N. poeticus* in high esteem, advising that: 'such a distinctly beautiful plant should be in every garden'.

This must have been music to the ears of Peter Barr. The Scot had done some investigating of his own and discovered a copy of John Parkinson's 1629 *Paradisi in Sole Paradisus Terrestris*, so he knew that many long-gone daffodil strains had once populated England's gardens. In his mind those puzzling daffodils of the past represented one fragment of the picture. The daffodils of the future were quite another.

By 1874, the pioneering daffodil hybridiser Edward Leeds was so infirm that the only way he could visit his beloved garden was by being wheeled there in his invalid chair. Word spread that he was readying himself for the worst.

Daffodil historians are divided about what happened next. Some suggest that Leeds launched a pre-emptive strike and really did compose a will stating that upon his death his unique collection — 169 *Narcissus* cultivars in the form of 24,223 flowering-sized bulbs — be put up for sale at £100, and that if nobody wanted to buy his life's work at that price everything should be burnt.

Another account suggests that Leeds forced the issue regarding the future of his unique collection and composed a letter to Peter Barr. In this scenario, outlined in the Kidderminster bulb seller Cartwright & Goodwin's 1908 pamphlet *The Latest*

Hobby: How to Raise Daffodils from Seed, Leeds issued Barr with an ultimatum. He told the Scot exactly how much he wanted for his daffodils (£100) and threatened to destroy them by 'digging them in' if Barr failed to buy.

Leeds knew that he did not have long. He offered his herbarium to Kew at no fee, which was swiftly accepted, but he left only one will that Joy Uings from England's Cheshire Gardens Trust could find while researching her fascinating 2003 University of Manchester dissertation on the topic. That will made no mention of anybody destroying Leeds's cherished daffodils, simply commanding that all his worldly possessions go to his widow, Ann. Leeds appeared to be a fundamentally generous man, Uings concluded after finding one acquaintance who described him as someone who 'would give you a share of anything he had whatever it cost him'.

Whatever the truth, two things are apparent. Leeds dearly wanted to find sanctuary for his daffodils with someone who appreciated their true value. Barr, in a hard-nosed commercial sense, did. Unfortunately he did not have enough cash. So Barr created a daffodil cartel.

On 21 April 1874 Barr wrote to E.H. Krelage & Son, one of the most celebrated nurseries in Holland, to inform them of the unique opportunity being presented by 'an old gentleman called Mr Leeds' who had created 'bicolors, majors, poeticus, incomparabilis, all shades from white to yellow and inter-mediate forms between incomparabilis and montanus and many other very unique crosses'. Grower Jan de Graaff reprinted the letter in the *1960 American Daffodil Yearbook*, revealing that Barr spelt out the situation's gravity very clearly.

> Mr. Polman Mooy [sic, a Dutch hybridiser] has entered his name as a subscriber of 10 guineas. We have done so too and three other gentleman amateurs have likewise put their names down for 10 guineas each.

Now as Narcissus are on the ascendant, shall we put down your name for 10 guineas? We believe it will be a very good speculation and another thing, we believe that, if the collection is not very soon bought, it will be destroyed, as the old man has put it in his will, if not sold before his death, it is to be destroyed. Drop us a line. Yours truly, P. Barr

E.H. Krelage & Son did not rise to the bait but Barr did manage to secure funds from two other Dutch parties — Polman-Mooy and Peter Van Velsen — and a bevy of English collectors: Herbert J. Adams, G.J. Braikenridge, W. Burnley Hume and the Reverend John Nelson, Barr's Norfolk-based brother-in-law.

Barr contributed enough cash to walk away with half of Leeds's daffodil collection, splitting the remainder between his fellow cartel members. When Nelson died in 1882 Barr took control of his Leeds daffodils, and by the time Van Velsen auctioned his share six years later (in the form of just over an acre of bulbs) its value had dramatically changed. Velsen's daffodils sold for around 15,000 guilders, a splendid sum given that a house in Amsterdam's prestigious Herengracht district changed hands for 10,000 guilders that year, and a wake-up call for the bulb market. Never before in Holland, the land of tulip-mania, had daffodils commanded such a price.

Frederick William Burbidge published *The Narcissus: Its History and Culture* in tandem with John Gilbert Baker in 1875, just a year after the Barr cartel buy-up of Leeds's daffodils, and it is clear from this book's beautifully illustrated pages that the Scot was moving fast. Barr had already begun growing Leeds's daffodil seedlings, wrote Burbidge, also noting just how tricky cultivating in England one of Parkinson's 'lost' varieties, the exotic 'White Hooped Petticoat Daffodil', was proving to be:

> This beautiful little gem seems to be rebellious to all the
> modes of cultivation ... Hot-bed, greenhouse, open air, all
> seem alike to fail. I saw last year at Messrs. Backhouse's
> nurseries at York 150 pots plunged into ashes, each
> containing a bulb, and amongst the whole I perceived one
> solitary leaf. Messrs. Barr and Sugden have imported the
> plant from Algeria by the thousand bulbs and I believe
> they have been equally unsuccessful.

Three years after selling his daffodils, Leeds finally, as Barr poetically put it, 'joined the majority'. The Scot had Leeds's bulbs but he needed something else: Leeds's horticultural notes detailing how he had created each distinct variety of daffodil. Without these nobody would know which two *Narcissus* strains had parented which 'child'.

Barr systematically started trying to put Leeds's collection in order. He cleaned, planted, cultivated, assessed and made efforts to identify each new bloom as, one by one, each bulb flowered; a painstaking process that took him a decade to complete. By the time he finished he had managed to secure the other great daffodil collection — another 192 fresh, original varieties — from the estate of William Backhouse.

Barr had no intention of repeating with Backhouse the mistake he had made with Leeds. He sweet-talked the late breeder's son Charles into letting him stay at Backhouse's home so he could rummage through his affairs in search of the precious hybridising notes.

'During the day we walked and talked amongst the daffodils, in the evening we searched amongst the departed's papers, which had not been destroyed,' Barr recounted, volunteering that he grilled Charles about his father: 'Was he of a nervous, sensitive, and gentle nature, and were his pursuits very refined?'

'He was all that,' answered Charles with some suspicion. 'Why do you ask these questions?'

'That was the conclusion I had arrived at from my study of his daffodils,' replied Barr, who would come to claim he could 'read' a person by studying the daffodil children that they spawned.

'A man in hybridising imparts the characteristics of his nature to the flowers he is raising,' declared Barr, brushing off Leeds's flowers as 'all more or less coarse'.

'I had little knowledge of the man,' he added dismissively, 'a few letters passed between us, but in the hurry of business one has no time to study handwriting. I therefore made some inquiries ... from those intimately acquainted with him, and found he was an off-handed, sharp businessman, with not much refinement. It is said he fertilised his flowers out of doors.'

Backhouse, who meticulously cross-pollinated his spring flowers under controlled indoor conditions, was clearly made of finer stuff. 'I am of opinion [Backhouse] saw in advance what the results of his work would be,' Barr said. 'Amongst the latter gentleman's daffodils I did not find one coarse flower.'

Years later Barr recanted, outlining to writer William Brockhurst how the intimate process of raising both men's daffodils had forced him to reconsider. 'Day after day, and year after year, I followed their work step by step ... till the work of these men came to be part and parcel of myself,' Barr explained. 'Edward Leeds was indeed a great man.'

The world's first Daffodil Conference began at the stroke of noon on Tuesday, 1 April 1884, at the Royal Horticultural Society's imposing large conservatory in South Kensington, London, with a clutch of heavyweights in attendance. Professor Michael Foster, a Cambridge physiologist, chaired the meeting, bringing down the hammer on a central order of business: a resolution proposed by botanist and lepidopterist Henry John Elwes and seconded by John Gilbert Baker, fellow botanist and Fellow of the Royal Society, designed to end centuries of narcissan confusion and give daffodils readily understandable names.

Garden varieties should 'be named or numbered in the manner adopted by Florists, and not in the manner adopted by Botanists', declared the proposal. A resounding 'Yes' from the assembled botanical dignitaries filled the room.

The event marks the beginning of the Narcissus and Tulip committee (later renamed the Daffodil and Tulip committee, and in 2013 replaced by the Royal Horticultural Society Bulb committee). From this moment on the Royal Horticultural Society would christen every new daffodil variety. Peter Barr considered it D Day; the date from which, as he put it, in an address to the monthly meeting of the Sea Point Horticultural Society in 1901, 'the fame of the daffodil was secured and like magic spread over the British Isles, extending to all Britain's colonies'.

To coincide with the conference an eye-popping exhibition hoped to convert a general public with no idea how mesmerising the daffodil could be. Gertrude Jekyll (1843–1932), the formidable garden designer and horticulturalist who understood what golden drifts of daffodils could add artistically to a garden, worked her magic. She conjured up flamboyant, artistically expressive displays of daffodils — in baskets, bottles and fan-like bunches — using blooms with tones that spanned the spectrum thus far created from sulphury white and pale primrose to a deep, glistening gold.

'The conservatory in the Society's gardens had a glow of warm yellow on either side,' noted an impressed attendee from *The Gardeners' Chronicle* in its 5 April review. The writer gushed about varieties 'that seemed to be but little known by any one out of the select circle of Narcissophiles', and said of Peter Barr and daffodils, ' He seems verily to have made them his.'

Within months of the Royal Horticultural Society's landmark ruling Barr was selling copies of a 1-shilling booklet called *Ye Narcissus or Daffodyl Flowre, and hys Roots* which listed 361 daffodils for sale under each one's official new name. These offerings promised to flower from January through to June and displayed blooms in shades as unexpected as reddish orange and even green.

Barr had called on Frederick Burbidge to co-write *Ye Narcissus*. An audacious example of what would today be called 'rebranding' it exhibits his capacity for masterful and speedy marketing. *Ye Narcissus* features old-fashioned scripts, eccentric spelling, illustrations 'Embellished with Manie Woodcuts', and presses home his vision of the daffodil as an intrinsic part of English cultural history. Letters supposedly dating back to 1610 are quoted, Narcissus myths retold and reference made to William Shakespeare — who had obligingly written of 'Daffodils / That come before the swallow dares, and / Take the winds of March with beauty' in Act Four of *The Winter's Tale* over 250 years earlier. Leeds and Backhouse are lauded as pioneering raisers of 'nearly all the new Daffodils'. The booklet concludes:

> We have said enough here to show how popular the Daffodil has ever been: It is essentially an English flower, full of vigorous grace ... of what is best and, most national in our character.

Barr used *Ye Narcissus* to outline the Royal Horticultural Society's new classification framework. It stated 'the name given to certain families of hybrid Daffodils — as Nelson's, Hume's, Barr's, and Burbidge's — are merely complimentary to these gentlemen for the conspicuous part they have taken in popularizing the Daffodil'. Barr and his cohorts had quite literally made the daffodil their own.

The Scot was not alone in thinking that the daffodil represented a potential gold mine. Across the water in Ireland, Cork nurseryman William Baylor Hartland (1836–1912) was busy scouring old Irish gardens for long-lost varieties of *Narcissus* to market and breed. In 1884 Hartland published the world's first daffodil-only catalogue, *A Little Book of Daffodils: Nearly 100 Varieties as Offered and Collected by W.B. Hartland*. Illustrated by Gertrude Hartland, the seedsman's niece, it inspired a new generation of daffodil lovers.

Two years later, England's first daffodil trade show took place at St Mary's in the Isles of Scilly, an archipelago 45 kilometres south-west of Cornwall with a milder climate than England and an earlier spring. *The Gardeners' Chronicle* said of the event, it 'speaks of difficulties overcome, it speaks of new prosperity for the Islanders, of the banishment of famine, of the establishment of industry and of the promotion of happiness'.

That autumn Barr published his updated *Narcissus* catalogue offering 'Daffodils, Daffadillies and Daffadowndillies'. He had dissolved his relationship with Edward Sugden and rebirthed his business as a family affair, partnering with his son Robert, who had trained at the Dutch Bulb Gardens of the Messrs de Graaff Brothers and was now running the Barr 'Experimental Grounds' in Tooting.

This new publication sang with Barr's vision. His daffodil was 'a cut flower ... most prized for ladies' dresses, bouquets, and vases [that] may be had in bloom from January to June', he wrote, providing thoughtful gardening tips on favourable locations for planting 'in Grass, Orchards, and by Streams and Lakes'.

He warned his public that he was far too busy to help them identify their stray daffodils, directed them to the Royal Horticultural Society instead, and quoted take-home prices for individual daffodil bulbs that ranged from 3 pence to 21 shillings — a fortnight's pay for the average 1880s labourer whose annual income amounted to £30.

Barr presented himself as a pioneering entrepreneur who had spent twenty years collecting varieties from as far afield as the Pyrenees. In reality, while he had begun exploring he knew he had much more to do. A river of gold in the form of commercially untapped *Narcissus* species flowed through the treacherous mountains of Europe and Barr wanted to find it.

The hunt was on.

CHAPTER 4

Barr's Dark Legacy

'Shrewd men of business have
sunk, and are still sinking, large
capital in these bulbs.'

Reverend George Herbert Engleheart,
1889

The annals of bulb hunting are laced with danger and romance. *The Perils of Plant Collecting*, an investigation into the hazards of the occupation by Edinburgh-based A.M Martin, a retired consultant physician and nephrologist, lists a maelstrom of mishaps endured by intrepid plant hunters who took it upon themselves to explore the wilds of North America and Asia between the seventeenth and twentieth centuries. Martin's catalogue of debacles ranges from being robbed, shot at, shipwrecked, disappearing without trace and dying in deserts, to contracting everything from frostbite, malaria, tuberculosis and syphilis to the bubonic plague.

These hardy souls needed financial backing, personal grit and plenty of time so it is perhaps no coincidence that Peter Barr took it up in earnest following the death in 1882 of his wife, Martha. I wish I could tell you that his experience fitted into the heroic mould cast by some of his pioneering forebears but this does not appear to be the case.

Barr had cultivated relationships with a number of useful Europeans including Alfred Wilby Tait (1847–1917), an English-born, Portugal-based wine exporter with an arrestingly aristocratic title (he was the Baron de Soutelinho) and a penchant for tracking down particularly elusive flowering plants.

In 1886 Tait published *Notes on the Narcissi of Portugal*, an otherwise earnest little essay in which he mischievously accused a coterie of daffodil lovers — 'Mr. C. Wolley Dod [the Reverend Charles Wolley-Dod], Mr. P. Barr, Professor Henriques of Coimbra, & Mr. Corder' — of infecting him with 'Narcissomania'. His investigations had convinced him that, contrary to common wisdom, more *Narcissus* existed in Portugal than Spain thanks to a historic redrawing of the national borders and he used *Notes* to list, in assiduous detail, each of the twenty-six different daffodils he had come across so far.

His write-up of 'N°13' is especially intriguing. A dwarfish dainty creature, deep sulphur in tone, it displayed delicate trumpets sculpted in the shape of tubes and petals that streamed back in the opposite direction. A Mr Edwin Johnson found it on a riverbank within striking distance of the city of Oporto – the first known sighting for around 250 years. Tait wanted Barr and botanist Charles Wolley-Dod (1826–1904) to pressure the authorities into having it named 'Henriques' in honour of fellow narcissophile Professor Júlio Augusto Henriques (1838–1928), Director of the Botanical Gardens at Portugal's University of Coimbra (a request that would fail). He made the point that this daffodil appeared in John Parkinson's *Theatrum Botanicum* under the name 'Cyclamineus' and advised 'the name must remain in abeyance until the roots I am sending to England flower'.

Barr knew that if he wanted to find this and Parkinson's other daffodils before

PETER HENDERSON & CO.

35 & 37 Cortlandt St. NEW YORK.

FOR AUTUMN PLANT-ING.

1892

BULBS, PLANTS, SEEDS.

his competitors did he had to get going. He targeted regions of promise, from the marshlands of Portugal to the Spanish Pyrenees, well aware that he was not alone. Other collectors such as Wolley-Dod, George Maw and H.E. Buxton had set their sights on the same exotic quarry.

What remains of Barr's travel journals — an edited, typed copy of the Scot's original writings — is held by the American Daffodil Library, and details his adventures by date, location, vehicle (often donkey) and species under pursuit. Barr describes travelling partners and the locals he cajoled into revealing where his elusive prey was hiding, and the subsequent traipses through remote, inhospitable landscapes from boggy fields and ancient tunnels to high, frozen peaks.

At times he travelled for miles without finding a thing. On other occasions daffodils seemed to be everywhere. Barr notes carefully which *Narcissus* is located where, harvests bulbs at every opportunity and packs them up for transport by steamer back home. His focus seldom strays far from his prize, as these diarised, oddly punctuated fragments show:

13.3.'87 Jose Marie, found in an orchard a yellow Nar.c
Ajax with perianth slightly lighter in colour than the
trumpet, trumpet with very thick tube, differing in this
respect from the yellow maximus ... Jose Marie, to make a
special journey to get more ...
18.3.'87 ... in the meadow in close damp soil, the large
yellow Narcissus Corbularia grows abundantly ... in some
cases flowers are very open. In some the petals stand
out at right angles, in others they are close up against
corona, as in Parkinsons' plate.
8.4.'87 ... On our return found that Jos had collected a
second basket of yellow ajax, in all 1000 to 2000 bulbs,
our haul for the day ...

10.4.'87 ... I found the head man ... who recognised the Ajax and came out and showed me where they were growing all about. They showed evidence of being wild, but having been thrown out when the land was being cultivated. It is clear that wild Narcissi will soon be extinct in this part of Spain ...

12.4.'87 Packed bulbs for London. Boxes 1 and 2 contain bulbs from Santa Maria and Reza about 2500. Boxes 3 and 4 contain about 1600 Narcis Ajax from the River Calvas 700 Ajax from Allarez and 300 Ajax from Venta de Soto Penedo.

One search for *N. Triandus* seemed all but fruitless, Barr reports. 'Mr Davidson informed me that George Maw got many from there. That may have been so but the evidence was wanting.' On another occasion, having bedded down for the night under the shelter of a rock ledge, woken at 4 am and fortified himself with bread and wine, he began the climb to 'narcissus quarters'.

'After a long and arduous trudge we came to the spot and found it cleared of Narcissi, 12,000 having been taken two days before. It appears that Célestin Passet and Pierre [celebrated French mountain guides] had had a commission. Very few remained and these will be collected.'

Barr recounted the experience of journeying with six Spaniards and a Pyrenean guide, 'sleeping under the rocks at night and travelling and collecting by day', in an address to the Sea Point Horticultural Society in 1901. Upon running short of food his entourage broke into a small cheese factory, helped themselves to whey and left 'money to cover our consumption'. That trick only worked once. On the way back Barr's posse tried to repeat the ploy only to discover that the owner had locked his factory up 'to keep us honest'. They bloody-mindedly broke in anyway, 'refreshed' themselves again, and went on their way 'rejoicing'.

The Scot demonstrated an equally uncompromising attitude to inns. Faced with a room with two beds in it and a charge that two travellers would have had to pay, Barr tucked himself into one bed for the first few hours then shifted into the other for the rest of his overnight stay.

Barr brought back to England some intoxicatingly beautiful daffodils. These included *Narcissus moschatus*, a miniature, silvery white, drooping trumpet daffodil found at 2,134 metres above sea level on Mount Perdu in the Spanish Pyrenees, *Narcissus asturiensis*, one of nine daffodils illustrated in Basilius Besler's 1613 illustrated work *Hortus Eystettensis*, and soulful, dwarf varieties dubbed 'Santa Maria' and 'Queen of Spain'.

He discovered a delicate pale yellow *N. triandrus* in north-west Spain on a blisteringly hot day. Spotting the wild species on a remote slope, he demanded that the young helper he had engaged for the trip, Angelo 'Angel' Gancedo, scramble up the rocks to harvest the bulbs. Gancedo obeyed but found it so challenging he started weeping with frustration. Barr would forever call the little daffodil 'Angel's Tears'.

Those who knew Barr developed a somewhat tongue-in-cheek attitude to him, Frederick William Burbidge archly telling one mutual friend that 'Peter the Great' was en route to Portugal: 'He says he starts fully armed and equipped this time and like the hero of your Cervantes [Don Quixote] he means business and adventure.' Burbidge also wrote to 'My dear Barr' himself, from Trinity College Botanical Gardens in Dublin, telling him in a letter dated 25 January 1888, 'I wish I was a rich man. I would take you to Portugal in my yacht and land you on the coast near your proposed raids.' Four months later Burbidge put pen to paper again to advise the daffodil hunter that visiting Montpellier in southern France might prove fruitful: 'This is a rich locality for Narcissi + perhaps worth your working.'

Barr appeared to be a hard man to miss. According to the Galician newspaper *El Correo Gallego*, which ran an article on 14 December 1888 headlined 'The Englishman of the Narcissi' (the name by which Barr was apparently known across Galicia and Asturia), he cut a bizarre figure even to locals used to the sartorial eccentricities of foreigners.

He sported a fur hat, big black boots, baggy knee-breeches and yellow leather leggings. His wide, collarless jacket looked more like a blouse than a coat, being 'crossed with multifarious pockets in every direction' and 'confined by a belt, the whole of a grey tone matching the beard and hair of the owner'. He filled his many pockets with paraphernalia, including field glasses and 'a multitude of pocket books, papers, and guides'.

Barr strode forth followed by 'a servant of as vulgar an aspect as can be found anywhere', noted the unimpressed correspondent, who dismissed the Scot's companion as a French mountain man 'whose chief characteristic is his knowing English hardly at all, Spanish even less, and his own mother-tongue only so far as the bad patois of the Landes and Western Pyrenees'. Barr's verbal style equally appalled the writer. The Scot spoke with the urgent speed of an express train, making it 'absolutely impossible to follow him' — yet that appeared to be the least of it.

Barr launched in with 'a torrent of questions, and a pulling out of note-books, and a jotting down in this place and in that, and requests for me to write down what he could not understand or what he could not spell in our language', the astonished reporter declared. 'And if by chance any matter, however out of the way, were touched upon, he instantly copied and made notes of it — and all this did not prevent his returning at once to the subject of paramount importance to Mr Barr at that period — namely — Narcissi!'

Barr explained he habitually spent months away from home in pursuit of 'information and botanical, geological and mineralogical specimens'. Once he had secured these he would return and divide his trophies among friends 'who devote themselves to each of these specialties'. What about the daffodil hunter himself, the reporter wanted to know: 'You do not take part in scientific review or publish ... work?'

'I am thinking about it and will keep you informed,' Barr fired back. The reporter did not believe a word of it.

'In Mr. Barr's work ... I failed to see ... any scientific object whatsoever, and I looked upon it as a species of monomania,' the correspondent wrote.

> But I now suspect it to have a very real commercial importance ... It is evident that [the daffodils'] high price in the market is quite a sufficient motive for the English to move heaven and earth and to try to acclimatize and reproduce them in their own country ... the sale of flowers, bulbs, etc., which goes on in the capital of the United Kingdom is something enormous.

Whoever wrote this article wanted to warn their compatriots about Peter Barr. Yes, he signalled a commercial possibility ('we have no lack of persons who are fond of cultivating and propagating flowers') but he also represented the end of the era of 'the ancient naturalists, those true martyrs to science, who, at the cost of infinite suffering — many perishing in the quest — opened out new horizons to the Geographical botany of the world with no other stimulus than glory'.

The American Daffodil Society's online archives feature a revealing series of excerpts from letters typed up at the end of Barr's edited diaries. One Portuguese seller appeared to have offered Barr '50,000 narcissus from Baroai' at a cost of '6/- per thousand'. Another missive from Wolley-Dod strikes a steely note of alarm.

Wolley-Dod had hunted down 'new' daffodils, having reintroduced *N. pallidiflorus* and *N. triandrus* var. *loiseleurii* to England in the early 1880s but Barr had him rattled. 'I should be sorry to hear that Peter Barr had become acquainted with any of your localities,' he wrote to an unidentified correspondent on 8 May 1885.

'If he can make it pay commercially, he will inevitably send out a collector and exterminate all the choice bulbs,' Wolley-Dod continued. 'He is very plausible, and wishes to know the precise locality for verification, but if he can see his way to making a few pounds by the transaction his promises of "Hands off" are worth no more than that of a Russian General. I speak from experience.'

Wolley-Dod went on to assure the recipient of his letter that, 'I did not of course give your name or the exact habitat. If I did, Barr would send over an agent and clear the whole species out, making a clear sweep.'

In one sense all this had happened before. Wild daffodils were taken from Europe's mountains during the sixteenth-, seventeenth- and early nineteenth-centuries by explorer–traders who introduced them to Britain and to the Netherlands. Yet Barr's systematic stripping of Spain and Portugal's wild *Narcissus* populations in the late 1800s would come to be seen by some botanical researchers as 'one of the darkest periods of plant exploitation', and at least one white daffodil from the Pyrenees, *Narcissus alpestris*, would be collected to the point of extinction.

What drove Barr was the dream that Victorian England would be swept away by the daffodil just as it had been by an almost fetishistic passion for orchids — a craze that spawned the delicious portmanteau word 'orchidelirium' and which had driven prices so high some London auction houses specialised in trading nothing else. It echoed a tulip craze two centuries earlier so frenzied that it almost crashed the entire Dutch economy, an episode Barr was well-acquainted with.

'During the tulip mania in Holland, a man exchanged a carriage and pair of grey horses, with all the trappings, and £300, for a tulip,' Barr told audiences during the 1901 South African leg of a world tour promoting what was, by then, 'the most popular flower of the day'.

'What the future of the daffodil craze may be I should not venture to forecast,' he mused. 'I have heard of a syndicate of amateurs who have given a daffodil raiser £500 for a portion of his new daffodils, and a trading firm who have invested £500 in the same man's new daffodils.'

What the Daffodil King did know was that his customers would pay increasingly high prices. By 1888 his catalogue quotes a fee of £5.5s for each bulb of 'Madame de Graaff', a sizeable, elegant, waxy white trumpet of considerable beauty. Narcissophiles with a less obvious vested interest echoed the salesman's confidence. The following year the Reverend George Herbert Engleheart (1851– 1936) announced in an essay for the *Journal of the Royal Horticultural Society* that he had no fear of a market 'bubble' in daffodil futures, and judged the flower's popularity 'unshakeable'.

'I am not of those who regard the extreme interest taken in Daffodils, and the immense demand for their flowers, as … a craze which will suddenly ebb away,' he wrote. 'The fact that shrewd men of business have sunk, and are still sinking, large capital in these bulbs [is] a guarantee that the Daffodil fashion will remain an abiding habit of springtime.'

Daffodil-mania lapped at the shores of Britain's colonies. 'To think it has taken all these years to render a daffodil fashionable,' mused New Yorker George H. Ellwanger (1848–1906) in his 1889 book *The Garden's Story*: 'As if a live flower were a ribbon, subject to the caprice of a milliner.'

Yet this is exactly what the daffodil had become and back in the motherland considerable efforts were being made to stabilise its reputation and capitalise on its star appeal. In 1892 Engleheart caused a stir by unveiling 'Golden Bell' a deep yellow Division 1 long-cupped dwarf trumpet bred from Backhouse's 'Emperor'. Such daring new narcissi clearly had the potential to become horticultural celebrities, however early public outings, such as Birmingham's first daffodil show, a two-day affair launched on Wednesday, 26 April 1893, at the Edgbaston Botanical

Gardens, could certainly have gone better. Fronted by William Hillhouse, secretary of the Birmingham Botanical and Horticultural Society (and a man who would predictably be dubbed the 'father of daffodil shows'), it consisted of fifteen classes, one judge (Burbidge) and barely any exhibitors — a situation for which its anguished organisers tersely blamed the unexpectedly early spring.

Barr made sure he was involved the following year, as did other influential daffodilians such as Engleheart, Jacob, de Graaff and Robert Sydenham, a successful Birmingham-based jeweller, bulb-seller, and founding member of the Midland Daffodil Society (later to become the Daffodil Society). Barr had focused his endeavours into propagation and by 1895 he was selling own-brand trumpet daffodils at 12 guineas a bulb. The average man's weekly wage then was little more than 1 guinea a day.

Barr wanted to step back from the day-to-day pressures of running his firm. On 26 October 1897, at the age of seventy-one, he became one of the first sixty recipients of the Royal Horticultural Society's Victoria Medal of Honour, an elite ensemble including botanical luminaries such as Sir Joseph Dalton Hooker, fellow *Narcissus* fans Burbidge, Charles Wolley-Dod and two remarkable women with a particular affection for 'Barr's flower'.

The first was the redoubtable Gertrude Jekyll, whose horticultural reputation had long been secure. Jekyll's surname was immortalised by the author Robert Louis Stevenson, a friend of the family, in his 1886 novella *Dr Jekyll and Mr Hyde*, and she already had a daffodil named after her, created by English grower John G. Nelson. Peter Barr proudly retailed the 'Gertrude Jekyll' as a striking, long-cupped trumpet 'almost uniform sulphur, very distinct'.

Ellen Ann Willmott (1858–1934), the second daffodil-related female Victoria Medal winner, was a woman of means who had bankrolled significant plant-hunting expeditions, as well as being an exceptional horticulturalist in her own right. Willmott had her extensive plant collection, reputed to feature more than 600 varieties of daffodil, tended by an army of gardeners at her home in Brentwood, Essex. Rumour had it that she booby-trapped her precious narcissi, installing trip wires attached to explosive air guns in an attempt to frighten off poachers.

Engleheart and Willmott knew each other. He had furnished her with *Narcissus* bulbs to assist her hybridisation efforts, and the year she took home her Victoria Medal he named one of his most attractive flowers, a Division 1 Trumpet, in her honour. 'Ellen Willmott' and 'Gertrude Jekyll' followed the fate of many such daffodil cultivars and after their fashionable flash they disappeared from the living rollcall of narcissi. If they still exist they are in hiding, and like the women they were named for they produced no known descendants to continue their genetic line.

As the twentieth century drew to a close the Midlands Daffodil Society's first show took place in Birmingham to showcase the flower's commercial potential. Liberal politician Joseph Chamberlain acted as president for the 1898 event, while John Charles 'J.C.' Williams, another business magnate and parliamentarian and who envisaged the daffodil becoming a valuable crop for the Midlands, held the post of vice-president.

Again virtually everything that could go wrong did. The heavens opened almost the moment the event began at Edgbaston Botanical Gardens as storms ripped across the skyline. Luncheon, held indoors, was a horribly over-crowded affair and the soggy conditions resulted in many boycotting the outdoor afternoon tea. At least one important exhibit arrived ruined, having been crated up disastrously poorly. The organisers immediately created a prize to be awarded to the best packed box.

Barr & Sons sponsored the show's inaugural Challenge Cup for the best daffodils in show — a field that could not have been thinner, featuring one lone exhibitor, the Reverend Joseph Jacob (1859–1926) from the North Shropshire town

Narcißus medio purpureus .
Narcißo .
Narciße .
Narciß .

19.

L. Narc.
lute
Ge. Narc
mida

of Whitchurch, whose pitiful efforts raised eyebrows rather than applause. The judges, trying to cobble together some skerrick of victory from the jaws of defeat, decided that Jacob should be recognised for at least bothering to enter and awarded him the trophy anyway.

Those assembled represented the largest group of growers that ever gathered in the United Kingdom; a narcissan *Who's Who* from both the trade and amateur fields, including Hogg and Robertson of Dublin, S.E. Bourne of Lincoln and, of course, representatives from Barr & Sons. Day two ushered in more miserable weather ('one continuous rain from morning to evening, which had a most disastrous effect upon the attendance', according to the official report) but at least everybody was indoors, attending a conference with Engleheart in charge.

Debate fastened on a suggestion that the organisation rename itself 'The National Daffodil Society' (it eventually become the Daffodil Society in 1963) and that a regular daffodil 'paper' be produced — a proposal doomed to fail as nobody wanted to take responsibility for it. Yet organisers hailed the event a success. Robert Sydenham (1848–1913), who saw the daffodil's international potential, explained his support in an interview published in the June 1900 issue of Birmingham's *Edgebastonia* newsletter. He argued that 'The Hollanders' had control of the bulb trade but that this could definitely change.

'I believe it will prove a remunerative pursuit in England,' he stated. 'There are vast tracts of land in this country ... where the soil is eminently suitable. I know several men who have taken up the business, and all of them have done well. One in particular was a schoolmaster in receipt of a salary of between £200 and £300 a year. He left school, took up bulbs, and is now making a great deal more out of them than he ever did in teaching the young ... how to shoot.'

Further south, where warmer temperatures ushered spring in early, the wonderfully named Thomas Algernon Smith-Dorrien-Smith (1846–1918), Lord Proprietor of the Isles of Scilly from 1872 to 1918, boosted the embryonic Cornish daffodil trade by seeking out and buying up large quantities of bulbs for his island tenants to cultivate. From these and other growers, avant-garde blooms continued to set hearts aflutter. When in 1898 Engleheart debuted his dazzling 'Will Scarlett', named after the youngest of Robin Hood's Merry Men, its restive, propeller-like creamy 'petals' and a large, ruffled, fiercely orange-scarlet cup provoked amazement and alarm.

This sensational Division 2 offspring of two wild species (seed parent *N. radiiflorus* var. *poetarum*, pollen parent *N. abscissus*) had people wondering just how far the daffodil could — and should — be pushed. 'Should a seedling ever attain to full red in both cup and petals, it would command an enormous price, but artistically considered, it would be somewhat of an outrage,' wrote a correspondent to *The Gardeners' Chronicle* (18 March 1899, p. 160).

Twentieth-century breeder Guy L. Wilson would blast 'Will Scarlett' but less for its showy looks than its genetic code. Wilson damned it as a 'shockingly bad flower [which] transmits many faults to successive generations' and the genetically ruinous parent of 'coarse and vulgar looking flowers'. Yet at 'Will Scarlett's' first showing Engleheart sold three of the only six bulbs he had bred so far to Kings Norton plantsman John Pope (1848–1918). The buyer paid the massive sum of £100. Never before in the daffodil's sphere had so much been paid for so little. Peter Barr must have wondered if his time had finally come.

His business did its utmost to maintain its killer edge, able to boast that of thirty award-winning nineteenth-century daffodil cultivars, all but one belonged to Barr & Sons. The man himself went walkabout after receiving his Victoria Medal, leaving England on a globe-trotting six-year jaunt — the first, and possibly the last, daffodil world tour. He gave talks under the banner-headline 'The Daffodil King'. The title had been jokingly bestowed upon him by botanist and *The Gardeners' Chronicle* editor Maxwell T. Masters (1833–1907), and Barr made sure it stuck.

Barr's lectures were thinly veiled sales pitches for his London-based business, revolving around the daffodil's rags to riches story and his role in resurrecting the fortune of this flower. 'I didn't make the daffodil, but I got possession of the only two collections in the world at that period, and I worked them into form, named, and classified them,' he said, using international celebrities to embellish his argument.

'I do not suppose that Oscar Wilde knew anything about daffodils, but there is no doubt in my mind that the great public are much indebted to him for the revolution in taste caused by his lectures on aesthetic colours,' Barr told audiences. 'He broke down the prejudice to yellow, the artists followed him, and the public followed the artists. About this time I had finished my work, and the collection represented about 500 distinct sorts ancient and modern.' Wilde singled out the sunflower but people soon tired of them, the Scot said. 'Then my daffodils came — and they came to stay.'

In May 1900, at New Zealand's Oamaru Public Gardens, Barr strode up to a man called James Gebbie Jr, asked him 'Are you the curator?' and pressed a business card into his hands. Gebbie, who had indeed created the gardens in 1889, told the local paper, the *Otago Witness*, on 3 May 1900, that he could barely believe his eyes.

'Is it possible that I am shaking hands with the Daffodil King?' Gebbie responded.

'Yes,' Barr grinned. The pair chatted for hours.

By now Barr had conquered America, Manila and Japan. Upcoming stops included South Africa, the Pacific Islands and Australia, where in the Victorian city of Bendigo on 23 November 1900 a local newspaper, the *Bendigo Advertiser*, reported that 'courtesies' were 'showered upon him' despite the fact that he bluntly dismissed the chrysanthemum, a local favourite, as a weed that anybody could grow.

'The daffodil wants handling,' Barr said, explaining that the daffodil had 'been in existence for about 300 years in Great Britain' but 'never was a people's flower

until 1884. Now it is seen in the poorest, as well as the richest homes, and it can be purchased for from 1/2d to £25.' He would go on to tell South African audiences that a single daffodil could be bought for 700 shillings, and that one woman had paid 1,000 shillings just to have one named after her.

'I created a new industry,' he declared. 'The street flower girls and hawkers of London do a profitable trade selling daffodils for at least three months in the spring. The leaseholder, or as he is called, "the King of the Isles of Scilly", and his tenants, derive a large annual income from the flowers I popularised. Some six years ago I chanced to meet Mr. Dorrien Smith, and he informed me he and his tenants had shipped that year to the mainland 150 tons of cut daffodil flowers.'

Early colonial settlers first introduced daffodils to Australasia but others had since become 'smitten', said Barr, who singled out the well-known George Sutton Titheradge (1848–1916), a Hampshire-born thespian who migrated to Australia and used his powers of oratory to broadcast the flower's virtues, as 'the exponent and special pleader of the family in his own fascinating style'.

'In the colonies of Australasia, especially in Melbourne,' Barr added, 'the flower boys sell an immense quantity of daffodils.'

Peter Barr ended his travels in 1903 and settled back in Scotland, concentrating his energies on a one-hectare garden at his home near Dunoon on the Cowal Peninsula and remaining a distinctive dresser to the end, typically attired in Harris Tweed trousers, a reefer jacket and a Glengarry bonnet. He died of heart failure on 17 September 1909 while visiting one of his sons in London, and is buried in Islington cemetery.

News of Barr's demise sparked obituaries around the globe. New Zealand's *Tuapeka Times* on 6 November 1909 observed that he 'had made himself rich and famous, and probably happy, by devoting nearly half a century to the study of one particular flower', and noted that when the 'Peter Barr variety', a daffodil named after him, first hit the market in 1900 it sold for a new world record of £50 a bulb.

When the 'Peter Barr' — a short, bold, solid, waxy white trumpet — had debuted it was heralded as unique. 'Never before has such a flower as this been submitted to the public gaze,' wrote E.H. Jenkins of Hampton Hill in *The Gardeners' Chronicle*, adding: 'The name Peter Barr (Daffodil King) is hardly likely to wane, and if so, this white-winged giant will certainly do its share in the perpetuation of the name.'

Reverend Joseph Jacob would come to wonder in his 1910 book *Daffodils ... with Eight Coloured Plates* what would have happened had the Scot not taken this flower in hand. Barr 'made the Daffodil. He travelled for it; he worked for it; he classified it; he advertised it,' wrote Jacob. 'The "King" as we like to call him, lived long and before he died he saw ... a race of new Daffodils appear ... like the sands of the sea-shore in number ... raised not only in our own islands but in lands beyond the sea.'

CHAPTER 5

Gold Rush

'From the upward tendency in the
price of new daffodils I should not be
surprised at them touching tulip prices.'

Peter Barr,

'Raising New Daffodils', *West Gippsland Gazette*, 2 October 1900

Now what, I hear you ask, has all this to do with my mother's garden? The answer
to that question shoulders its way through the soil every spring as my mother
waits and watches, never entirely sure which flower will show its face next. 'Sir
Watkin' is a classic whose appearance always delights. With soft, lemony petals
that blend prettily into an orange fluted cup, this large Division 2a first emerged
at some point prior to 1868 and is attributed to William Pickstone, born in 1823
to a Lancashire family so poor that he had to earn a living from the age of seven
as a handloom weaver. As a child Pickstone got to know an older neighbour, John
Horsefield (c. 1792–1854), another working class weaver. Horsefield was a lean,
spare, intelligent man with such passion for all things botanical he memorised Carl

Linnaeus's plant catalogue system by pinning it to his loom so that he could study it as he worked. He served as Prestwich Botanical Society's president for thirty-two years during a time when — thanks to Linnaeus's system, which allowed anybody who could read and write to identify any plant — botanical research became as open to the poor as to the educated elite.

Working class enthusiasts would meet in pubs to share their hard-won discoveries and toast new specimens with rounds of ale and song. By the mid-1800s these egalitarian societies disbanded as science migrated back beyond the reach of the impoverished to be taught exclusively in academic institutions. As Angela Jean Walther mapped out in her 2012 Iowa State University graduate thesis *From Fields of Labor to Fields of Science: The Working Class Poet in the Nineteenth Century*, Horsefield's demise coincided with the end of a remarkable time.

Shortly after Horsefield's death a jaunty dwarf daffodil appeared in his garden, presumably bred by him. Named *Narcissus* 'Horsfieldii' in his honour (the 'e' is believed to have been omitted by accident) it fast became a money-spinning favourite, as big a hit with the buying public as William Backhouse's 'Emperor'.

William Pickstone would have heard of Horsefield's death and his valuable, posthumous daffodil. Pickstone had done well and left his working class roots far behind. Displaying dogged determination and an entrepreneurial eye he parlayed an apprenticeship at a Manchester warehouse into a lucrative manufacturing business of his own. He earned enough money to retreat at the age of forty-three to Wales, where he started investigating 'mining proclivities', as he revealed in *The Gardeners' Chronicle* on 29 December 1894. Here, beside a tributary of the River Dovey beneath the shadow of the mountainous Cywarch Crag, Pickstone 'met ... the great Welsh daffodil now known to florists all over the world as "Sir Watkin"'.

The way Pickstone recounted the tale, he nourished his sweet-smelling foundling for years, taking bulbs with him when he moved away first to Hampshire then back to Wales. Around 1880 a canny businessman spotted them flowering on his land, and offered to buy them. Pickstone agreed and watched the harvested cut flowers sold off in Liverpool and Manchester by a merchant desperate to keep the secret of their location to himself.

So startling was this daffodil that if Pickstone wore one in his lapel he would be stopped in the street. He sent specimens to Peter Barr but agreed to let the Welsh horticulturalist James Dickson, the first to arrive with a good offer, buy 1,000 of his bulbs.

In 1884 William Brockbank visited from Manchester to see the flower for himself, find out from where it hailed (Pickstone refused to tell him) and inform Pickstone that 'your daffodil' had already been registered with the Royal Horticultural Society as 'James Dickson'. Brockbank told Pickstone that he, Pickstone, 'ought to have the naming of it'.

'I said that mine was too obscure a name for so beautiful and so sweet a flower,' Pickstone recalled, prompting Brockbank to suggest it be called 'Sir Watkin' after the prominent Welsh politician Sir Watkin Williams-Wynn (1820–1885). Given that Williams-Wynn 'is the uncrowned King of Wales and had farms near the place of its birth, I readily accepted the suggestion', said Pickstone who sent the politician a box of the handsome flowers. Within forty-eight hours Sir Watkin and his wife Lady Williams-Wynn had agreed to 'became sponsors to the unnamed Welsh baby'.

Whether James Dickson cared about the name change is unknown. His priority was cornering the market on this valuable beast. He beat Barr and all other potential competitors to buy up around 23,000 'Sir Watkin' bulbs for close to £1,000 — a new record as far as anyone knew for a stock of any daffodil cultivar. The public adored it. By 1910 it ranked as one of the most expensive daffodils available, selling in England for 3 shillings and sixpence a bulb, though by then the Dutch had control of it, growing more in the commercial bulb fields of Katwijk alone than one visiting British narcissophile thought 'were to be found in the world'.

'Sir Watkin', like Wales's attractive Tenby Daffodil, *N. obvallaris*, was virtually exterminated from its native location by avaricious bulb dealers and it left some experts scratching their heads — including, ironically, Brockback who announced in 1894 that 'Sir Watkin' was no newcomer, advising that it had been identified long ago as Narcissus *incomparabilis*.

'Strange indeed that it should have been altogether forgotten and overlooked for 250 years at least,' Brockbank noted in the *Chronicle* with no little suspicion, 'and stranger still that Dickson should have become the purchaser of this novelty when all the time it abounded within a few miles of their Newtown nurseries, quite unknown to them, and where it had flourished for at least 100 years.'

If there is a moral to this tale it is that there is nothing simple about any of these daffodils' stories, just as there is actually nothing simple about life. As 'Sir Watkin' prompted Brockbank to realise; as my mother understands with every new *Narcissus* face that unfolds each spring; as I learned that day in Sydney when confronting an unexpected cancer diagnosis, there is often far more going on beneath the surface of any situation than ever first appears.

The nineteenth century had ushered in a bewildering array of wonders. Bulb seller William Baylor Hartland composed a lengthy poem about 'Our Starlit Era' for his 1897 daffodil catalogue that doubled as an homage to Queen Victoria's Jubilee, and paid tribute to a number of these inventions:

> ... The Telephone! The Telegraph!
> The Phonogram! The Photograph!
> Bright Röntgen rays that lift the veil,
> To peep through hearts, where ills prevail ...

He did not mention the daffodil but he could have. Peter Barr probably would have. As the millennium ticked over, the Daffodil King's reign came to an end, yet this

flower's modern renaissance had just begun, as author Arthur Martin Kirby made clear in the 1907 hardback *Daffodils — Narcissus and How To Grow Them*, America's first homegrown book on the topic.

Kirby's dainty green offering on the 'daffodil craze' provides historical titbits, gardening advice and ethereal black-and-white photographs. It informs readers that 'scores of rival enthusiasts in Europe and Great Britain' not only breed new daffodils but are capable of creating bulbs that can change hands for as much as US$100 apiece.

'More than that, there are some daffodils that may never be seen by the outside world,' Kirby breathed. 'A coterie of six wealthy daffodil lovers in England buys up the bulbs of any new varieties of exceptional beauty and merit ... paying extravagant prices for the sole ownership of the coveted beauties.'

This secretive clique would happily drop US$2,000 on just five or six bulbs and close ranks to protect their high-priced treasures, revealed Kirby. 'One of the compacts of this close club is that at the demise of any member, his or her bulbs are to be distributed among the remaining members of the monopolistic band,' he wrote of the 'Daffodil Six'.

Less elusive than these elite collectors were the efforts being made to thoroughly democratise this flower. In 1910 the Reverend William Wilks (1843–1923), a British-born, Oxford-educated, plant-loving parson and Secretary to the Royal Horticultural Society, constructed an impassioned argument for *Narcissus* in his preface to the Reverend Joseph Jacob's book *Daffodils ... with Eight Coloured Plates*.

Its virtues simply abounded, wrote Wilks observing that the daffodil had proven itself unique in 'beauty and general utility ... so cheap and so easily grown, it blooms so early as the harbinger of spring, it travels so well, and it keeps fresh so long in water'.

Wilks had experience of hybridising various flowers and made his horticultural name by creating the Shirley Poppy, an electrifying new multi-coloured poppy bred from wild field flowers. He felt sure the public would embrace *Narcissus* 'until there will not be a garden in the land that has not a little patch devoted to Daffodil'; and would have known, just as any experienced cultivator did, that occasionally something quite unexpected might appear.

'Fortune' sprang from nowhere if Somerset bulb trader Walter Thomas Ware (1855–1917) can be believed. The son of T.S. Ware, a well-established Tottenham plantsman, Walter helped his father secure a near monopoly of *Narcissus poeticus* var. *ornatus* in the 1880s by paying French suppliers 18 francs per thousand bulbs, a fee that covered bulb quantities up to a million. Walter Ware grew the bulbs to feed the cut bloom trade, a livelihood so lucrative that twelve stems of 'Emperor' and 'Sir Watkin' could sell for half a crown. He bulk-supplied bulbs to British and Dutch dealers, also dealing in crowd-pleasers, such as the magnificent shell pink Darwin tulip 'Clara Butt', named after the Bristol-born opera singer. Ware cornered the market there, too, and his decision to offload his huge stock of 'Clara Butt' bulbs resulted in the flower plummeting in price.

Ware never claimed to have bred the 'Fortune' bulb, insisting to his grave that it spontaneously flowered early in the spring of 1916 in a seedling bed at his property in Inglescombe near Bath. The moment the sixty-one-year-old saw it he knew he had a winner on his hands. Striking in stature and form 'it resembled a giant golden *incomparabilis* with a snub-nosed corona that shone with the most opulent orange anybody had ever seen'.

Its timing could hardly have been more germane. World War I had left the flower industry in tatters. That January brought the *Military Service Act*, which introduced conscription for single men aged sixteen to forty-one, and by May a second Act extended the British war machine's reach to married men. It left few

at home to cultivate blooms and a shrunken market for flowers. As the Reverend Joseph Jacob put it in the 23 September edition of *The Garden*, 'The noiseless tenor of the way of the Daffodil has been rudely interrupted.'

Daffodil shows had become skeletons of their former selves. 'If in these times of anxiety and sadness we feel justified in having any flower shows at all, no one will grudge our great modern flower of spring a place among them,' Jacob mused forlornly of the previous year's Royal Horticultural Society event. '[For] one half day, at any rate, we were in the pleasant land of Daffodildom, with our cares and troubles left behind.'

At the Society's 1916 show things had gone from bad to worse. 'Nothing from Watts, nothing from Guy Wilson, to name but [two] of those from whom a yearly supply of "good things" may in normal years be counted upon,' Jacobs continued, adding soberly, 'Time claims his victims no less than war.' Engleheart was unwell and Ware was ailing. Even so Ware made his way to the Midlands Show with a selection of twelve cut daffodil stems, including the eye-catching 'Fortune', to enter the Bourne Cup.

'Fortune' dazzled an unnamed reviewer from *The Garden*, whose 6 May report declared it 'a wonderful bloom ... the last word in red-cupped giant incomparabilises'. Jacob was no less impressed.

'The red-cupped "Fortune" shown by Mr. Walter Ware in his fine Bourne Cup twelve at Birmingham was one of the most outstanding flowers of the year,' he wrote. 'It will assuredly be heard of again.'

Ware seemed clueless on the question of his flower's parentage, though his manager J. Firman hazarded the guess that it could have been 'Sir Watkin' crossed with 'Blackwell'. The episode left some suspicious. That Ware had acted as a bulb agent for Ellen Ann Willmott prompted modern-day *Narcissus* expert Noel Kingsbury to speculate in his 2013 book *Daffodil* that Willmott could have bred 'Fortune', and Ware — inadvertently, perhaps — taken credit for her work.

Whatever the truth this new bloom's future hung in the balance. Disease was ripping through daffodil populations from England to Holland, leaving plants stunted, discoloured and lethally frail. Dutch grower Matthew Zandbergen, then a youngster, saw first-hand how badly affected his father's and uncle's *Narcissus* stocks — the source of the family's livelihood — became. 'I remember helping to lift, during growth, a badly infested and newly acquired stock of "Sir Watkin",' he wrote decades later. 'The bulbs were carted into barges and taken to a factory to be converted into starch during the 1914–18 war.'

Few growers seemed untouched and all were rattled. The future appeared dire. Some believed the pest was a fungal blight called Fusarium. Others argued the culprit was stem and bulb eelworm, a scourge first identified in 1858. Frederick William Burbidge had written to Peter Barr in 1889 after a devastating 'mysterious disease' had attacked some '*jonquille* roots'. Twelve years later it swept through the garden of Percival Dacres (P.D.) Williams (1865–1935), resulting in the loss of years of work.

By 1909 the affliction had reached the grounds of the Somerset breeder Alexander (Alec) M. Wilson (1868–1953), who described it as 'a veritable Black Death'. Every remedy Wilson tried failed. 'My stocks were valued for Income Tax purposes at £12,000 and two years later I had not £200 worth left,' he wrote. 'So, in the words of Rudyard Kipling: "I saw the things I gave my life to broken" and "had to stoop to build them up with worn out tools".' The disease reached Ware's flowerbeds shortly after he discovered 'Fortune', destroying around 300,000 'Horace' daffodils alone with terrifying speed. Ware had just four precious 'Fortune' bulbs in his possession and took drastic action, seeking out far-flung, hopefully uninfected gardens to send them to. They included the Scottish estate belonging to Major Ian Ashley Moreton Brodie (1868–1943), or 'the Brodie', as he was known.

The Brodie, 24th laird of Brodie Castle (his Scottish ancestral home in Morayshire), was a prolific and dedicated hybridiser. Accounts vary over whether he bought one bulb of 'Fortune' or two for £80 but either way he planted the daffodil within a specially walled area of his gardens and, with fierce dedication, got to work.

Of all the hybridisers that had come before, none seemed to match his scope, precision and ruthlessness. It would take a seasoned soldier to safeguard 'Fortune' at a time when *Narcissus* lovers were fighting their own version of a daffodil world war.

Schooled at the aristocratic Eton College in Berkshire before joining the Scots Guards, the Brodie first experimented with cross-breeding forty-nine varieties of daffodil in 1899, the year he left to fight in the Boer War. By 1902 the wounded soldier was back, and upon seeing that his crosses had born fruit, hybridised another 375 new plants. World War I prompted another bout of active service culminating in the Brodie returning home in July 1917, and the arrival of 'Fortune' at Brodie Castle.

The Brodie hybridised tens of thousands of new daffodils during his half-century career, of which over 400 would ultimately be judged good enough to earn names. He sourced the best bulbs available, experimented with a discipline that is tempting to call military and kept meticulous accounts of parent plants, time of year and even the weather. 'His immaculate exactness when it came to planting daffodil seeds, each three inches apart in rows eight inches apart, reflected his precision training,' wrote enthusiast Margaret 'Meg' Yerger in the June 1980 edition of the American Daffodil Society's *The Daffodil Journal*.

'When the blooms came out they made such a uniform display they might have been likened to a proud military unit lined up for review. And reviewed they were! Unless a flower was of a high standard demanded by Brodie, it was removed as fast as if it had been picked off by a sniper.' He systematically destroyed the vast majority of seedlings he created, flowers he considered inferior, by burning them.

By the time 'Fortune' reached the Brodie, the Royal Horticultural Society's

Twee. Span. noua, or upa which means flower
water, holy.

Narcissus and Tulip committee was in crisis mode. Deeply rattled by the carnage taking place it appealed for help from the Society's horticultural experts at their Surrey research grounds Wisley Gardens, where they had a new £20,000, state-of-the-art laboratory to play with. A Dr Keeble was the researcher in charge and as Joseph Jacob rather archly put it, the daffodilians had given him and his staff 'something to do ... Blessings upon [them] ... should they develop into St. Georges and slay this menacing dragon.'

The Narcissus committee was waging war on another front, fighting to stop the impending closure of the loss-making *Daffodil Yearbook*. Enthusiasts from as far afield as Australia and New Zealand mobilised to protest the looming calamity. 'By the rank and file of amateur gardeners in America the golden sails of this loveliest fleet of flowers were but just hailed on the horizon, and now our binoculars, so to say, are snatched from us,' wrote Mrs Francis King from Michigan to *The Gardener's Chronicle*. Another correspondent named only as 'Narcisso-phile' observed darkly that 'A peculiar futility seems to dog the footsteps of the Daffodil'.

Things moved fast at Wisley Gardens, where twenty-six-year-old Bradford-born research student James Kirkham Ramsbottom (1891–1925) diagnosed the daffodil blight as a parasitic nematode, or eelworm, called *Tylenchus devastatrix* (now *Ditylenchus dipsaci*) with such speed that by 1917 he had begun trialling a cure. Ramsbottom proposed soaking bulbs for two to four hours in hot water baths, a procedure that killed the eelworm yet left the plant unharmed. His breakthrough was so effective that half a century later Matthew Zandbergen observed in the 1967 *Daffodil Year Book* (yes, it survived) that Ramsbottom's principle still held true.

It 'has practically not been amended', wrote Zandbergen, 'except that his middle range temperature 114–115°F [45–46°C] for four hours has since been found to give more satisfactory results'.

With Ramsbottom's help, P.D. Williams saved around 2,000 daffodil varieties, noted Zandbergen, whose father had acted as Williams's agent and sold the relieved breeder's bulbs to Dutch growers at 'fantastic prices'. The deceptively simple cure had far-reaching implications, effectively securing a new global era of daffodil trade. In 1924, amid mutterings that there had never been a more worthy recipient, the Royal Horticultural Society awarded him the Peter Barr Memorial Cup. On 9 February the following year, while visiting New York on a lecture tour, the thirty-three-year-old plunged from a high hotel window and died.

'Fortune' fared better. Although Walter Ware would never know it (he died shortly after attempting to find infestation-free homes for his bulbs), his swift action saved the daffodil. It also left the Brodie in possession of the only uninfected bulbs. The Scot proceeded to corner the market so effectively, many mistakenly believed it one of his own creations. The Brodie considered 'Fortune' such a prize he kept a black-and-white photograph of it in his private collection, which his son Ninian showed to historian Margaret Woodward in 1996. She told me she recognised the subject instantly.

The image, which has since gone missing, showed a young male gardener grinning at the camera, dressed in dungarees and proudly clutching a large, single daffodil bulb. Below it someone had written in pencil, or possibly in ink which had since faded away to a smoky grey: 'This is worth more than my year's wages.' Beneath that another hand had scribbled '£25'.

'Fortune' became a darling of the cut flower market. It also proved to be fertile, paving the way for a new battalion of experimental, orange-tinged daffodils. The Brodie deployed it and other sought-after blooms to fund his breeding program, each spring issuing orders that his employees pick, stack and box the best buds before sending them by train (the laird's castle had its own station) to Edinburgh, Glasgow and south of the border to Covent Garden in London.

His bulbs sold worldwide, but closer to home the Brodie's own fortunes took a complicated turn. He and his wife Violet had three sons — the classic heir and two spares. David, the oldest, succumbed to diphtheria in 1911. Michael, the next in line, died after a traffic accident in 1937, which left Montague Ninian Alexander Brodie (1912–2003), who was in the process of establishing a career as an actor, to inherit the Scottish estate.

In 1939 Ninian married fellow thespian Helena Budgen. Hollywood star Stewart Granger made a cameo appearance at the wedding as their best man, but what most interested the press was that the groom rushed straight from the wedding to Streatham Hill Theatre to appear in a matinee performance of *Romeo and Juliet*.

In 1943 Ninian fathered twins and became laird when his hybridising father passed away. Ninian's mother and a gardener called Mr Milne continued to work with the Brodie's daffodils but Ninian had little aptitude for it, admitting to one radio interviewer: 'I've got no green fingers at all.'

The Brodie's daffodils bloomed each spring but the family's luck had turned. Their crumbling castle, its contents and sprawling estate became such a financial drain that six years after Helena's death in 1978 Ninian sold the lot to the National Trust (via the Secretary of State for Scotland) for £130,000, on the condition that he could continue living in a small apartment on the property.

Two years later the Trust opened up the property to the paying public (Ninian took a particular delight in guiding visitors through and developed a reputation for reciting risqué limericks during tours). A breeding program, led by retired gardener Leslie Forbes, began to re-create the Brodie's daffodil collection, now recognised as internationally significant, and by 1988 enough cultivars had been reclaimed to win the Brodie daffodils National Plant Collection status. Yet the drama continued. As Noel Kingsbury related in *Daffodils*, Forbes had labelled 156 cultivars but the soil in which they grew had not been mapped and holidaying children destroyed all the labels. The work had to begin over again.

Brodie's guardians kept searching and in 2004 Brodie Castle head gardener David Wheeler received two bulbs of 'Perth', a yellow and orange daffodil believed

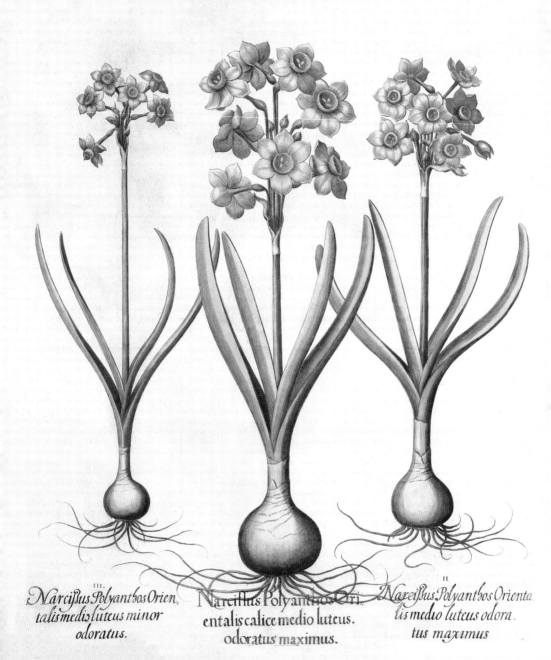

III.
Narcissus Polyanthos Orien,
talis medio luteus minor
odoratus.

Narcissus Polyanthos Ori,
entalis calice medio luteus.
odoratus maximus.

II.
Narcissus Polyanthos Orienta.
lis medio luteus odora.
tus maximus

extinct, that the Brodie had registered in 1929. John A. Hunter, a well-known breeder in New Zealand, had sent it, advising that his father had bought it during the 1930s.

In 2011 the Trust commissioned a husband and wife team of daffodil experts to assess the Brodie collection: Duncan Donald (former curator of the Chelsea Physic Garden in London) and Kate Donald (a former International Daffodil Registrar of the Royal Horticultural Society). The pair, who run a business called Croft 16 Daffodils near Poolewe in north-west Scotland, has long been tracking down pre–1930s daffodils and when asked by Noel Kingsbury why they picked that date as the cut-off point for heritage varieties, Duncan explained that this was around the point at which 'Fortune'-derived varieties began becoming important.

'We don't like orange cups,' Duncan told Kingsbury who recorded the conversation on his blog, 'they all look like each other, we don't care for the "Fortune" look.'

In 2015 long-held plans to sell Brodie daffodil bulbs to the public became a reality. The National Trust's ultimate aim is to locate and recover as many of the Brodie's missing 'soldiers' as possible but for now emphasis lies with conserving the Major's existing regiments which put on a glorious show each spring. At the last count, over 200 different daffodil cultivars grow in the castle grounds.

'Fortune' also flowers in my mother's garden, and her notes describe it thus: 'Division 2a Y/Y. A large flower on a twisted stem. Strong yellow perianths (petals) and orangey yellow corona (cup) that is very bright. Highly thought of.'

On the whole she is right.

CHAPTER 6

The Hybridiser's Tale

'A young aspirant to fame asked me if it were possible to get a red Daffodil. I have seen one, but it was in a dream.'

Peter Barr,
The Daffodil Yearbook, 1915

There is probably no more daffodil-like daffodil than the Division 1a cracker called 'King Alfred', a plant so robust it has dwelt in my mother's garden for at least the better part of a century. It has a 'large flower on tall stem, large trumpet rolled over at the edge', as my mother observes in her notes, and can lay claim to being one of the best-known daffodils ever grown. As American Daffodil Society founding member George S. Lee Jr pointed out in the Society's 1966 *Daffodil Handbook*, sixty-seven years after this daffodil's debut, it remained the most widely grown variety.

'Without question, the creation of King Alfred ... was the greatest single advance ever made in the progress of daffodils,' Lee wrote. 'Those who think there

is only one daffodil — the yellow trumpet seen in florists' windows — have King Alfred in mind.'

Upon its debut in 1899 this radiant flower immediately won over the Royal Horticultural Society's Narcissus committee with its charisma, size, regal bearing and richly uniform gold tone. That year happened to be the millennial anniversary of the Anglo-Saxon King Alfred the Great's death, hence its name. Percy Kendall, a grower from Devon, brought the flower to the committee, but Percy had not bred this flower; his father, John Kendall, a daffodil enthusiast and former solicitor who died nine years earlier, had. John was well known in the daffodil world and for the committee members, seeing 'King Alfred' must have been like meeting a ghost. The committee awarded the flower its highest honour, a First Class Certificate, and Peter Barr leapt upon it. Within a year his company was retailing its bulbs for a hefty 6 guineas apiece.

This flower's parentage remained a mystery. Cognoscenti deduced it was most likely the outcome of a cross between two widely admired yellow trumpets — the intensely toned *Narcissus maximus*, a pre-1576 variety with artfully twisted sepals, and either 'Emperor' or 'Golden Spur'. But a third character had had a hand in the development of this flower: Walter Hill, a talented gardener who worked for Percy (John Kendall's son had owned a market garden and a florist) and lived at 'The Garden', a rented cottage in the village of Newton Poppleford. As Rose Brady revealed in an investigation for the Daffodil Society that celebrated the 100th birthday of 'King Alfred', Hill had selected it from all the other seedlings, and raised it himself.

After Percy's death in 1910, Hill cultivated 'King Alfred' in earnest. He dug in around 8 hectares of the bulb, employing locals to speed-harvest it and stack the bunches of budded stems into his 'glory hole', a shack behind his cottage, where warmth from a stove coaxed the buds into flower.

Hill had around 2,880 daffodils packed each day, loaded onto horse-drawn carts and taken to the village railway station in time for a London train, en route to Covent Garden's flower market. Ada Grigg, who was born in 1903 and worked for Hill, told Rose Brady she earned sixpence an hour packing cut bunches which were in high demand. 'The "King Alfred" came out just after the Cornish crop had finished, and we used to pack them in blue tissue paper, three dozen to the box,' she recalled.

Hill used horses to plough up the bulbs every July. Grigg and her fellow workers then had to scour the fields for bulbs, standing them up and turning them repeatedly until they had dried out. The best would be sold, the rest replanted, and 'King Alfred' put Newton Poppleford on the map.

Hill died in 1935. In 1999 the elderly Grigg, still a resident in the village, unveiled a commemorative blue plaque, echoing the daffodils' wrapping paper, on the wall of her former daffodil boss's home. It read: 'On March 14th 1899 the King Alfred Daffodil which was bred on these premises by Walter Hill, nurseryman to Percy Kendall, received a First Class Certificate from the Royal Horticultural Society.'

Despite Hill's work it is hybridiser John Kendall whose name remains most closely associated with 'King Alfred', the ancestor of an estimated 90 per cent of all yellow trumpet daffodils since bred. Ironically, in all likelihood its creator never saw it bloom.

It is not money that usually drives those wanting to engineer the daffodil. Given the general rate of return upon effort, that is probably just as well. For those captured by the desire to perfect this flower, however, there is no more prestigious award than the Royal Horticultural Society's Engleheart Cup introduced in 1913 to honour the best twelve daffodil cultivars raised by an exhibitor, as demonstrated with one clean stem of each. The award is named after the Reverend George

Herbert Engleheart, a man who became known in *Narcissus* circles as the 'Daffodil Maker' upon breeding hundreds of exceptional new varieties. In so doing he changed the daffodil, and gardens like my mother's, for good.

Engleheart, who declared that it is as difficult to create a good daffodil as it is to breed a racehorse, shared a distant genetic connection to 'the Dean', William Herbert. Born in 1851 on Guernsey, one of the Channel Islands, Engleheart studied at Oxford University's Exeter College, became curate of Leicester in 1877, married Mary Isabel Evans the following year, and experienced fatherhood within twelve months of having wed. By 1881 he had relocated to a vicarage in Wiltshire and his daffodil adventures began.

Engleheart saw how much money the shrewd men of business, as he called them, were sinking into the daffodil and suspected its popularity would endure. An erudite man, he eloquently anthropomorphised his flowers. Some had a tendency to melt 'in sun and wind', he wrote in his 1889 'Seedling Daffodils' article for the *Journal of the Royal Horticultural Society*, yet others struck him as 'better-behaved'. One 'droops like limp muslin on a scorching day', while another 'is both pale and precocious'. Of a seedling that has yet to flower, about which he knows little, he pleads that it will 'not die suddenly, as [if a] pale precocious child'.

Engleheart's first 'hit' arrived in 1892 with 'Golden Bell', a graceful, nodding dwarf trumpet that won him a First Class Certificate from the Royal Horticultural Society, as did his 'Ellen Willmott' five years later, named in tribute to his celebrated acquaintance. In 1898 he stopped the Birmingham Show in its tracks with 'Will Scarlett'.

Engleheart befriended John Charles Williams, a Cornish-born politician who viewed daffodil farming as a potential solution to the depressed economies of Cornwall and the Midlands. Williams's vision fired up Joseph Chamberlain to such an extent that the politician appears in one official portrait sporting a daffodil in his buttonhole instead of the orchid he customarily wore.

Williams presided over the 1897 Daffodil Spring Flower Show in Truro, an event designed to expose the flower to the public, which featured Thomas Algernon Smith-Dorrien-Smith of Scilly as vice-president and Engleheart as judge. The daffodil-spruiking Chamberlain–Williams double act continued the following year in Birmingham at the first Midlands Daffodil Show, where Chamberlain served as president and Williams as vice-president. Williams's cousin, Percival (P.D.) Williams, also got in on the act. Both visited Engleheart's gardens to buy bulbs — fuel for a breeding war that would involve years of fierce competition as the cousins pitted daffodil again daffodil at subsequent spring flower shows.

In 1900 Engleheart received the Victoria Medal of Honour from the Royal Horticultural Society and *The Gardeners' Chronicle* — which had declared in its 28 April 1900 issue that 'Everybody is, for the time being, Narcissus mad' — singled out just two individuals for commendation: 'Barr, the Daffodil King, the pioneer of the present fashion, and Engleheart, who has done so much for which the dry-as-dust scientists will thank him as heartily as do the aesthetes.'

The following year Engleheart bought Little Clarendon, a run-down property in the Wiltshire village of Dinton with 11 hectares. He had made massive improvements to the poeticus daffodils in particular, creating 'varieties blooming all through the season', as Reverend Wilks wrote in the preface to *Daffodils ... with Eight Coloured Plates*. 'He has given us, too, flowers with good stiff petals, full of substance, and able to maintain a really flat appearance.'

Engleheart continued to impress. In 1923 he offered up 'Beersheba', a stately, sizeable Division 1 daffodil with a long ivory trumpet and pale, radiant petals. Engleheart probably created it during the traumatic height of World War I and, though nobody knows for certain, most likely named it after the 1917 Battle of Beersheba.

In 1933 the *Daffodil Yearbook* featured a dedication to Engleheart. He 'is alone and outstanding as the producer of the most beautiful hybrid narcisii', wrote

P.D. Williams, adding sadly that a devastating 'eel-worm and fly' infestation meant Engleheart's breeding days were over, and that he had turned his attentions instead to issues such as 'the investigation and excavation of Stonehenge'.

'Engleheart is a highly-cultured man, well read and most appreciative of everything that is beautiful,' said Williams. 'His handwriting, at the age of eighty-two, is still the neatest and most perfect I have ever seen.'

Given his predilection for writing letters to *The Times* newspaper, that was probably just as well. Engleheart variously demanded the urgent protection of Stonehenge (from vandals intent on chiselling their names into the ancient stones), pleaded for a proper analysis of streamwater quality, urged people to collect rain-water for irrigation, and offered handy hints on how to outfox pickpockets ('with a large & strong safety-pin').

His reputation as a daffodil man continued to grow. Tasmania's *The Mercury* newspaper ran an article on 10 January 1934 headlined 'The Flower Garden', which observed that while great progress had been made in the daffodil world, 'there has been no more important or far-reaching break than that obtained by the Rev. G.H. Engleheart when he successfully crossed *Poeticus ornatus* with the best forms of *Narcissus tazetta* ... The result of this cross was an important race of *Narcissi* with the hardiness and fragrance of *ornatus* and the free-flowering properties of the *tazetta* type.'

On 14 March 1936 *The Times* published a piece by Engleheart on the topic of Stonehenge. He died the following day but his daffodils live on. Today's specialist bulb sellers stock some of his most impressive offspring, a coterie that includes 'Bath's Flame' (a long-time resident of my mother's garden), 'Seagull', 'White Lady' and 'Beersheba', a flower whose descendants range in name from 'Adri' to 'Zest' bred by a Who's Who of twentieth-century daffodil makers. That list includes Alec Wilson in Wales, Australia's William Jackson, England's Douglas Blanchard, Gerritsen & Son in the Netherlands, Murray Evans in the United States and Guy

L. Wilson from Northern Ireland, one of the most exceptional breeders of them all, who described 'Beersheba' as a daffodil of 'arresting beauty and outstanding purity'. Some years ago my mother decided she could not do without it and added half a dozen bulbs to her collection.

The impulse to create new daffodils is a curious one. Hybridisers freely admit that what drives them, the unquenchable urge to create the perfect *Narcissus*, is fundamentally fraught. Harold Koopowitz, Professor Emeritus of Biology at University of California at Irvine (UCI), president of the American Daffodil Society and hybridiser par excellence, described the classic, elusive, flawless daffodil to me as 'a totally unnatural thing' when we spoke in 2015. It is 'round and flat,' he said. 'Nature abhors that.' The South African-born Koopowitz is a rare breed with an unusual, and unusually informed, perspective. Britain's nineteenth-century daffodil hunters stripped hundreds of thousands of *Narcissus* bulbs from remote Mediterranean populations with little apparent concern for the future, and when I asked Koopowitz what effect he thought Peter Barr had exacted on wild daffodil populations his reaction was blunt.

'Probably decimated many of them,' he replied, adding, 'That said, the question is not straightforward. It is really very complicated. One has gut feelings but one's gut feelings are usually not correct.'

Koopowitz, like Barr, spent years tracking down Portuguese and Spanish daffodils but his motivations differed. By the late 1970s this scientist had become the first director of UCI's Arboretum, a botanical garden and research centre, and as his career progressed he witnessed a growing awakening to the global impact of habitat devastation, a halt being called for in the unregulated harvesting of wild plants for sale, and the effect that gene banks for endangered species (such as that launched by the UCI Arboretum) could have.

Koopowitz found himself puzzled by why, when a group of plants became threatened with extinction, some seemed to fare better than others. He investigated orchids, a specialty of his, in Zimbabwe until political events intervened. 'The forest I was in was owned by a white farmer and ... was taken away from him,' he explains. 'So I looked around and thought, "Why don't I do this with daffodils?"'

In 1998 Koopowitz embarked on his first Spanish daffodil field trip to compare the sexual tactics used by abundant daffodils with types that were rare. He and his colleagues delved into herbarium records to deduce where particular daffodils had been discovered in the distant past, and used the historical information to attempt to locate them anew. He walked through valleys shimmering with waves of narcissi — 'literally a million daffodils in flower' — and concluded that the likes of Peter Barr seemed to have done little damage. In other locations nothing could be found at all. Yet 'Some populations "wink out" and disappear naturally,' he counters, 'So if something is not there you can't be sure if it was collected out or not.' Additionally, daffodil flowering was sometimes linked to fire. Heavy rain occurring the year after a burn-out could generate an ocean of daffodils in a location that had previously, apparently, held none.

Koopowitz suspected scarcity hinged on the biology of the species. Some commonplace *Narcissus* created more seeds than rare varieties, and could mature up to five years faster. 'I learned a lot there,' he says of that first field trip. 'I think it must have been around about the time I started breeding the miniatures.'

Plant-crossing had always appealed to the scientist. 'That is where the sciences and the arts get together; you use the science to make your hybrid but your aesthetics is what decides whether or not it is worthwhile,' he says. 'But there's been a century and a half of daffodil breeding and many of these flowers are pretty close to perfection. There aren't many more improvements that you can make [and] for me the main challenge is to do what no one else has done before. Miniatures were the orphans in the daffodil world.'

In the mid-twentieth century Cornish hybridiser Alec Gray (1895–1986) made the sphere of miniature *Narcissus* breeding his own. Gray became a world authority on the subject, spending decades breeding dozens of new versions of these delicate little gems. He joined the Royal Horticultural Society's Narcissus committee in 1938, took home the Peter Barr Memorial Cup for his work on dwarf hybrids in 1945, and was awarded the American Daffodil Society's Gold Medal in 1966.

From Gray's pedigree stable, three particularly noteworthy multi-flowered siblings emerged — 'Tête-à-Tête' (registered in 1949), 'Jumblie' (1952) and 'Quince' (1953). Their father was unknown, while their mother was what Gray described as an aberration — 'the one fertile flower of "Cyclataz" that I have ever come across in some 30 years of growing that plant'. 'Freaks like this do occur,' he reflected in his 1965 lecture 'Small Hybrid Daffodils'. 'If the number of plants of any one sort grown runs into thousands, the chance break may happen, be spotted and propagated.' 'Cyclataz' is a miniature daffodil that blooms with two or three orangey-yellow flowers on each stem. Portugal's Alfred Wilby Tait created it at some point before 1922 by cross-breeding 'Soleil d'Or', another jaunty poly-floral, with the wild miniature *N. cyclamineus*. The moniker 'Tête-à-Tête' translates from the French as 'head to head' (appropriate for a bunch-flowered *Narcissus*) and is a pun, of sorts, on its 'grandfather' Tait's name.

Six years after creating his little star Gray was selling it for 5 shillings a bulb. It became the most popular miniature daffodil ever bred, with Dutch growers dominating its production, selling bulbs by the million. Reliable, faithful, and one of the earlier daffodils to flower in my mother's garden, it is also very tough. In 2014 'Tête-à-Tête' and two other popular daffodils, 'Dutch Master' and 'Ice Follies', were found to be highly salt-tolerant, making them potentially useful plants for saline-affected ground. Gray said he earned relatively little from 'Tête-à-Tête' although it did, in its way, make him immortal, as every bulb is a clone of the one he first grew. In 2011 DNA testing finally cracked the mystery of its parentage

when scientists established genetic evidence of two species — *N. cyclamineus* and *N. tazetta* — and raised the possibility that a new, equally lucrative daffodil could be created by attempting to replicate the original cross.

After Alec Gray retired, amateur breeders struggled with dwarf daffodils, never the easiest of hybridising realms. As American breeder Mrs James Birchfield from Ashburn, Virginia, complained in *The Daffodil Bulletin* of February 1962, 'While working on miniatures I would find it additionally helpful to have a built-in magnifying glass and be able to work while standing on my head.'

From what Koopowitz could see, breeders were creating the same thing time and again, flowers that he judged 'cute but fairly blah, yellow or white and they either looked like a small jonquilla or little trumpet'. Nobody seemed to be attempting Division II daffodils, 'those split coronas which some people don't like,' he says. 'I thought, "Gee I reckon I can make these, and fill in a gap".'

The scientist sought out new varieties to hybridise with as he travelled through Spain and Portugal, selecting the tiniest flowers with the most seductive characteristics and, unlike Peter Barr, collecting seed rather than uprooting bulbs. He exploited previously unmined genes in fresh wild species — *Narcissus* never before used for breeding — and experimented, even tapping the American Daffodil Society's research fund to investigate daffodil 'embryo rescue', an in-vitro cultivation technique involving growing fertilised ovules in a synthetic medium, test-tube baby style, in an effort to overcome incompatibilities between two parent plants. 'The results were dreadful,' he laughs. 'It was one of the most ugly distorted flowers I have ever seen.'

Koopowitz's biggest 'hit' so far is his debonair miniature 'Itsy Bitsy Splitsy', so-named because it is a Division II split corona, it is small and the three words rhyme. Florida Daffodil Society president Linda van Beck told me she suspects it may come to rival the success of Alec Gray's champion, but Koopowitz disagrees.

'Everybody does want the next 'Tête-à-Tête' but I don't think I have it yet,' he says, stressing that though talent agents for bulbs exist in his home state of California ('Where else?' he laughs) he is not in it for the money. Quite apart from anything else, daffodil breeding is a numbers game: 'The more seeds you get the more chances you have of winning. But even under carefully controlled conditions they do as they damn well please. It is really just gambling.'

Koopowitz's scientific background has given him a hybridising advantage — 'a feel for the genetics' as he puts it — as have the lessons learned on that first Spanish field trip: 'Things I hadn't really thought about; that there are lots of daffodil species that bloom in the autumn, not just the spring.'

Botanists have long known of autumn-blooming daffodils. John Parkinson wrote about 'the greene Autumne Jonquilla' and in 1753 Carl Linnaeus identified *Narcissus serotinus*, a minuscule creature that sports a startling array of white tepals and one of the tiniest yellow coronas ever seen. Classified as a Division 13 wild variant, *N. serotinus* blooms after hot dry Mediterranean summers, grows wild from Portugal and Turkey to Israel and North Africa, and issues a deep, forceful perfume detectable even at night.

Koopowitz experimented with *N. serotinus* and *N. viridiflorus*, another autumn-flowering daffodil from Morocco. This star-like Division 13 jonquilla is nocturnal; a ghost-like, semi-skeletal being that releases its scent after dark to lure pollinating moths. It is green from tip to toe (*viridi* translates from Latin as 'green', *florus* as 'flowered') and American garden writer Louise Beebe Wilder (1878–1938) thought it seriously creepy, describing it as a 'strange spidery, almost evil-looking creature'. Under natural circumstances it could never mate with a spring-flowering daffodil.

Yet *N. viridiflorus* has galvanised an elite group of modern-day breeders: from Australia's Lawrence Trevanion to the late Californian Manuel Lima, a shy figure who recognised its potential in the 1960s. Bob Spotts, a past president of the American Daffodil Society, inherited Lima's seedlings and felt his approach was as noteworthy as his floral achievements in that he seldom had much money but was very good at persuading fellow enthusiasts to help him with his breeding stock. Lima repaid such collegiate generosity with pollen, seed and bulbs.

Koopowitz's work with *N. viridiflorus* has prompted some to suggest that autumn daffodil shows be organised and others to wonder whether his work could revolutionise the entire daffodil industry, because some *N. viridiflorus* hybrids have flowers that can last up to a month. 'Embedding that [characteristic] into a big showy daffodil is going to take a lot of work,' Koopowitz counters. 'I don't think I'm going to get involved in that.'

N. viridiflorus has also inspired New Zealand hybridiser John A. Hunter, who has used it extensively in his quest to create cutting-edge daffodils. Hunter first exhibited daffodils at the age of nine, in 1945, and is a mine of information, telling me that New Zealand's first daffodils — varieties such as the yellow double 'Telamonius Plenus' (known by the likes of my mother as 'Van Sion') — were imported by white settlers in the early 1840s. Hunter's great-grandfather grew daffodils in the 1880s for Henry Budden (1842–1902), a botanical artist and one of the country's first nurserymen to specialise in bulbous plants, and his family has grown them ever since.

Hunter began raising daffodils, or 'the damn things' as he occasionally calls them, in 1949 — 'Longer than anyone else,' he says firmly — across almost every daffodil division, and was the first person awarded the world's three major *Narcissus* advancement awards: New Zealand's National Daffodil Society's Gold Medal in 1997, the American Daffodil Society's Gold Medal in 2011 and the Royal Horticultural Society's Peter Barr Memorial Cup in 2012.

Even as a thirteen-year-old Hunter kept careful notes in his schoolbook, detailing the precise pedigree of his first eleven crosses. He always systematically divided up and labelled everything. 'It's a lot of extra work,' he says, describing how successful fertilisation results in pods which may contain over fifty seeds but that those seeds, once planted, take around five years to flower. Even then patience is

pivotal. Only after several years of flowering can a new daffodil's real quality be properly judged. To Hunter, 'It's been intriguing ... a marvellous experiment in genetics.'

By the 1960s Hunter was making 120 crosses a year, producing thousands of new seedlings every season and cremating those not up to scratch: 'You've got to be brutal ... You can't grow them all.' *N. viridiflorus* appealed because it underwent a genetic change very early in its historical life which 'doubled its chromosomes from 14 to 28', Hunter explains. 'The modern daffodil has a chromosome count of 28. I thought if I could get hold of that I'll cross it with the modern hybrids and see what happens.' He squeezed his contacts to obtain some pollen, isolated it within a capsule, kept it in his deep freeze until the New Zealand spring month of September when daffodils start to flower and then used it to fertilise specially chosen flowers. In 1999 he registered one particularly innovative new daffodil with the name 'Emerald Sea'. 'It is green and white — opening green and going after two or three days to white and green,' he observes with satisfaction. 'That one has probably gained me more fame than anything else.'

During six decades of hybridising this breeder has seen daffodil colours change. 'Pink and yellows had just started to come into appearance in the late 1950s but before that, that type of flower was virtually unknown,' he says. 'Even what we call a pastel pink, that didn't appear in England until around 1910.'

The way Hunter tells it, a freak poeticus bloomed with an orangey-red corona tens of thousands of years ago in the European mountains, and it is to this one flower that every other brilliantly coloured daffodil owes its hue. 'Daffodils around about 1910 would have been maybe six or seven generations away from the species and you need that variation before you would get those colour changes,' he deliberates. 'The daffodils of today would be nearer twenty generations away from the species so there's a huge genetic diversity within them. That makes them a beauty to breed with; the variation that you can get.'

Hunter has named daffodils after everything, including US newsreader Walter Cronkite, but would never name one for himself. 'I remember going along to a show and someone telling me proudly, "The Americans have raised a nice daffodil and named it after me",' he says. 'I told them it wasn't a good idea. You're on the top of the heap for about two or three years then you are superseded and down at the dump.'

He sees himself as standing on the shoulders of devoted *Narcissus* breeders from the past. 'The daffodil has been one of the most well-documented flowers for pedigrees, it is possible in some cases to trace the pedigrees back to the 1840s,' he explains. 'No raiser can claim that any new great seedling they have raised is solely their own work as every hybridiser relies on all who went before them for the cultivars for their breeding stock. Every grower's good new seedlings are but stepping stones for further advancement. It is only when working with species crossed with species that a raiser can claim it to be their own work.'

Asked what fuels his passion for daffodil breeding, he replies, 'The thrill. Every year walking out and seeing the new varieties flowering for the first time, there's always something entirely different.'

Hunter treasures a letter from Guy L. Wilson which arrived after he wrote to the Irish breeder to let him know that he had crossed an old English flower called 'Royalist' with one of Wilson's yellow trumpets, 'Ulster Prince'. Hunter never met Wilson, though he reveals that a friend did and described him as feisty: 'An old bachelor that had given his whole life to the flower, I suppose one could accept that.'

So what will happen to Hunter's own daffodil collection. The breeder replies by saying he understands exactly why Edward Leeds threatened to destroy his hard-won narcissi and that he suspects his own work will go 'the way of every other good raiser's collection, to absolute disrepair'.

'My wife has threatened to build a big fire out the back, put me on the top when I die and all the bulbs with me; and my two cats,' he tells me. 'We've been married over fifty-five years and all through that time she's given me immense help with them. I could not have achieved in the last half century what I have done without her.

'We sit out there when they're in flower, a chair on each side of the bed, looking at all the new seedlings, evaluating them all, putting a peg beside the ones that we want marked, noting in our plan book where these flowers are.'

Breeders may change the daffodil but in Hunter's opinion perfection will never be attained. 'These days we have "marvellous cerise red-pinks daffodils",' he muses.

'If I could bring my mother or grandfather back to life they wouldn't believe the daffodils I've got now. They would think you had faked them because they look so, so unreal. Those pinks are now turning to mauves and if one concentrated on those particular ones, and kept crossing them together …' and the elderly hybridiser pauses to consider what he has done and what he has left to do.

'If I had another lifetime in front of me I believe I could turn these mauve-tinted daffodils into blue.'

The
Daffodil
Fairy

CHAPTER 7

A Daffodil Code

'Of all floral catalogues, a daffodil
catalogue is the most exquisitely
tantalizing. The further you
read, the deeper the gold.'

George H. Ellwanger,
The Garden's Story, 1889

In the ancient species *Narcissus tazetta*, myth, meaning and wishful thinking
collide — although you would not know it upon encountering 'Soleil d'Or',
probably the oldest cultivar growing in my mother's garden. Dated by botanists
to before 1731, 'Soleil d'Or' is one of spring's earlier offerings. It is tall, wide-
leaved and bunch-flowered, displaying up to eight yellow blooms, sporting
feisty orange coronas, on each of its slender green stems. With character and
joyful pizzazz it so delights my mother that she recently decided to augment her
collection with an extra fifty bulbs.

'Soleil d'Or' is a true survivor. It is adaptable, persistent, hard to kill and easy to love — a classic tazetta, from a clan which possesses such fertility that its name now denotes an entire division of cultivars (Division 8), characterised in part by having up to twenty flowers on each stem. Some varieties, as my mother's notes point out, have an unmistakeably heady aroma. Others radiate a perfume so subtle that it is imperceptible to people yet deeply enchanting to honeybees.

To humans *Narcissus tazetta* has simply always been there. It has journeyed far from its original home in south-western Europe and gathered many names on the way, from Polyanthus Narcissus, Nosegay Daffodils, Paperwhites, Suisen (by the Japanese), Joss Flower to Chinese Sacred Lily in China, where it is considered auspicious.

As the nineteenth century drew to a close it certainly brought daffodil lovers such as Peter Barr some unexpected good fortune, word of which even reached those in America on Friday 12 July 1889, when the weekly journal *Science* hit the newsstands.

A remarkable Egyptian exhibition was about to open at London's Royal Society with help from the head of Kew's Royal Botanic Gardens, *Science* revealed to its readers. This show would feature botanical wonders discovered by archaeologist William Matthew Flinders Petrie (1853–1942), the thirty-five-year-old grandson of scientist and cartographer Matthew Flinders, whose 1801–1803 circumnavigation of Australia had generated international acclaim.

At a time when Ancient Egypt thrilled the public (rich Victorians even hosted mummy-unwrapping parties), Petrie seemed as adventurous as his celebrated forebear, having returned from his 1888 expedition to an ancient cemetery in the Egyptian settlement of Hawara with what appeared to be a stunning haul. His team had unearthed everything from long-abandoned workmen's tools, loose coins and ceremonial papyrus sandals to antique children's toys. They discovered ancient coffins and mummies, some floating in water alongside dismembered, bobbing skulls.

Included among the finds were 'wreaths ... found in wooden coffins, either resting on the heads or surrounding the bodies of the mummies', *Science* relayed, explaining that Petrie had identified certain plants. Items named included: 'cones of papyrus pith', 'chrysanthemum flowers', 'rose petals', 'scarlet berries of the woody nightshade' and, most interestingly of all, in two mummy garlands, 'Polyanthus narcissus (*N. tazetta*)'.

Petrie believed that some of these mummified bodies had been adorned with wreaths that included *Narcissus* in what he called — in *Hawara, Biahmu, and Arsinoe*, his 1899 book of his adventures — 'a wonderful state of preservation'. These daffodils looked to him like 'modern specimens' but what were they doing here? 'This plant was probably introduced from Palestine, where it now occurs in great abundance,' he concluded.

Petrie stood as a celebrity to the Victorian public, a golden figure in this glorious age of discovery. His life story dazzled with unexpected drama. The son of an engineer with no formal education to speak of, he had become obsessed with Egypt at the age of thirteen after reading a book on the topic. In adulthood he became a polymath who spoke six languages, revolutionised the modus operandi of collecting archaeological artifacts, was granted the nation's first chair in Egyptology (at University College London), founded the organisation that would become the British School of Archaeology in Egypt, and made arguably more major finds in his field than anybody else ever would.

His dedication to science was such that upon his death in 1942 he ensured his fully fleshed head, and the brain residing within it, was delivered to the Royal College of Surgeons of London, where it apparently remains to this day.

Yet Petrie never linked *Narcissus* directly with Ancient Egypt. He believed the *Narcissus* wreaths were Greek, and he made that clear. Peter Barr knew this. 'In 1888 during some excavations in the cemetery at Hawara, in Egypt, some floral wreaths were found,' Barr told a South African audience during his daffodil world

tour, 'one of which consisted of the bunch flowered daffodil, and is supposed to have been made by a Greek artist resident in Egypt about the first century before the Christian era.'

Yet a myth arose linking the daffodil with the captivating romance of the ancient Pharoahs. To this day, an often repeated daffodil 'fact' is that Ramses II, a legendarily all-powerful Egyptian leader who died in 1213 BC, was ceremonially prepared for the underworld by having *Narcissus tazetta* bulbs placed on his eyes or around his neck. *Narcissus tazetta* has been linked to Ramses II by the identification of dry, scale-like fragments from the plant on the outside of his mummified remains, yet the Egyptologists I spoke to struggled with the notion that daffodils held any meaning for Ancient Egyptians.

Dr Linda Evans from the Department of Ancient History at Sydney's Macquarie University considers it unlikely that the daffodil was known in Ancient Egypt as she has never seen it mentioned as a flower with religious or symbolic importance, unlike the lotus or the waterlily, and Petrie's Hawara site dated to the Roman period. Christopher Eyre, Professor of Egyptology at the University of Liverpool in England, is equally cautious, advising that the daffodil does not come to mind as an Egyptian symbol. Lise Manniche, Assistant Professor of Egyptology at the University of Copenhagen and the author of a series of books on the subject, counselled me that she had never encountered evidence of *Narcissus* in her research, either in botanical remains or in representations.

It is easy to understand why some promoting the cult of the daffodil latched onto the cache of Ancient Egypt with such fervour. Some contemporary mythmakers in the cosmetics industry are attempting to do something similar, embracing the supposedly remarkable properties of *Narcissus tazetta* bulb extract. The promises are legion, including guarding against the stresses of the modern-day environment, warding off wrinkles, bolstering the skin's elasticity, combating unwanted hair growth and even, on a cellular level, arresting the passage of time.

The notion that *Narcissus tazetta* could hold the secret to eternal youth is seductive but highly unlikely. What is clear is that to some people this plant spells 'profit', which is perhaps surprising given that few other garden flowers in history have been so damned by association.

The classic daffodil is yellow, a colour long linked with decadence, scandal, decay, vanity, madness, jaundice and other forms of sickness. Yellow was considered emblematic of disdain during the Middle Ages, when prostitutes, lepers and Jews were at times forced to badge themselves by wearing it, an act repeated by the Nazis in the run up to World War II.

This colour has the power to turn otherwise neutral phrases nasty — think 'yellow journalism' (meaning sensationalised media coverage) and 'yellow peril' (the offensive term relating to Asian immigration) — so it is little wonder that the daffodil fared poorly when the charming concept of 'floriography', a language 'spoken' with flowers, emerged in early nineteenth-century Europe.

French writer Charlotte de la Tour triggered the craze for floriography dictionaries with her captivating 1819 tome *Le Langage des Fleurs*. Fast translated into English, Tour's book claimed to divulge the secrets of ancient floral traditions that included the historically dubious Turkish practice of women in harems using a mysterious code comprised of flowers to spell out secret messages to their lovers. The Georgian, and then the Victorian, public adored the notion that they could bypass social etiquette by speaking to each other with individual blooms, tussie-mussies (little posies) and 'talking' bouquets.

Suddenly flowers could be used to flirt, insult, abuse and dismiss — as long as both the sender and recipient understood what they represented. A flurry of code-breaking floral dictionaries appeared, enabling these covert communiqués to be translated with precision.

From the outset the daffodil had trouble. La Tour lists the meaning of narcissus as '*égoïsme*' (selfishness) and asphodel as the more sinister '*Mes regret vous suivent*

au tombeau' (my regrets follow you to the grave). Jonquil fared somewhat better, representing feelings of *désir* (desire).

Boston-born poet Frances Sargent Osgood, as notorious for her scandalous liaison with Edgar Allan Poe as for her subversive literary output, dug the daffodil in deeper in her 1841 handbook *The Poetry of Flowers and Flowers of Poetry*. She stated that jonquils represented desire, daffodils spelt out 'deceitful hope', False Narcissus (*Narcissus pseudonarcissus*) meant 'delusive hope', and the Poet's Narcissus translated as 'egotism'.

The floriography fad swept through nineteenth-century popular culture, leaving ever-changing nuances, contradictions and absurdities rippling in its wake. Over a century later Brent Elliott from the Royal Horticultural Society's Lindley Library became so irritated by modern-day inquiries about what individual flowers 'meant' (nothing more than people wanted them to mean, as far as he was concerned) that he compiled first a card catalogue then an Excel spreadsheet of different definitions culled from Lindley Library's historical books.

According to Elliott's findings the meanings for jonquil and asphodel altered little over time but the daffodil's fate deteriorated so badly that a Victorian lady presented with a single bloom could read into it anything from unrequited love to deceitful hope, folly or disdain. Peter Barr and his ilk rose to the challenge. By the time they had finished rebranding this flower, it had shed many of its less unsavoury associations and sported instead an image revolving around its traditional Catholic meanings of rebirth and hope.

The dramatic rewriting of folk history stuck, but every now and again a poignant reminder of the daffodil's difficult past would re-emerge, as it did in Warsaw after a man called Marek Edelman let it be known that since at least the 1960s he had received a bunch of yellow flowers, including *Narcissus*, every 19 April from someone who never revealed their name.

The date marked the anniversary of the Warsaw Ghetto Uprising in 1943, the most famous Jewish revolt against the Nazis during World War II, in which Edelman had played a key role. Prompted to use the daffodil as a silent message of tribute, Edelman and his compatriots established a tradition of honouring the fallen by placing daffodils at key sites including the city's Monument to the Ghetto Heroes every year on that day. When Edelman died in 2009 the ceremony continued. Well-wishers put daffodils on his grave in Warsaw's Jewish ceremony, as the flower had come to symbolise the Uprising, representing remembrance and respect. An official Daffodil Campaign was launched in 2013 to mark the event's seventieth anniversary.

On 19 April 2013 Warsaw's church bells rang as sirens wailed and volunteers handed out yellow daffodil badges to remind people of the yellow Star of David the Nazis had forced the Jews to wear. Jonathan Ornstein, executive director of Krakow's Jewish Community Centre described seeing so many people wearing the daffodil emblem as 'one of the most moving things I've seen in all my years in Poland'.

That year street artist Dariusz Paczkowski unveiled a mural in the city depicting Edelman and a single daffodil flower. Meanwhile, elsewhere in Europe, a storm was beginning to brew around two much older daffodil artworks.

Their story began on Friday, 6 January 2012, when horticulturalist John Grimshaw, Director of the Yorkshire Arboretum, made a visit to London's National Gallery to see the 'Leonardo da Vinci, Painter at the Court of Milan' exhibition, billed as the most complete collection ever held of the artist's work. On show together, for the first time, were two versions of da Vinci's artwork *Virgin on the Rocks*, one of which belonged to the National Gallery, the other to the Louvre in Paris.

These images are magnificent — intense, complex, highly detailed and almost identical. Each features the Madonna and another woman with two children in a steep landscape of rock and plants. The French-held image is the earlier of the two, dated from 1483 to 1486 and considered groundbreaking because of the almost scientific accuracy with which da Vinci depicted the natural world. The second painting is believed to have been completed in 1490. Prior to their joint appearance National Gallery director Dr Nicholas Penny predicted 'that the experience of seeing these masterpieces juxtaposed will be one that none of us will ever forget'.

That proved true for John Grimshaw, who visited to examine the exquisite botanical elements of da Vinci's work. At first glance the two images looked similar, 'but there are subtle differences in the imagery of the figures and their settings, and to me there is a surprising — if not shocking — difference in the plants in the landscape,' he relayed in his blog. 'In the French painting the plants are beautifully rendered, with the detail one expects from the great botanical artist that Leonardo was: an *Iris*, *Polemonium* and *Aquilegia* are clearly recognisable.

'Replacing the *Iris* in the London version is a clump of apparent *Narcissus tazetta* — but it is no normal daffodil. The flowers are good enough, but they arise on bracteose scapes, from a clump of plantain-like leaves.'

To Grimshaw the daffodil's blooms were accurate but the rest of the plant simply wrong, meaning that the artist had created an unreal, Frankenstein creature that seemed to go against everything da Vinci usually did. Puzzled, the horticulturalist surmised that da Vinci had 'invented his own flowers for Paradise'. With an apparent shrug, he noted, 'It is nice to think that he envisaged daffodils there.' Ann Pizzorusso, a geologist and Renaissance art historian, harnessed Grimshaw's observation while researching her 2014 book *Tweeting da Vinci*. Pizzorusso argued that the Louvre version of *Virgin on the Rocks* could be viewed as a superb example of da Vinci's genius but that the botanical inaccuracies of the National Gallery's version made it profoundly troubling. To her it seemed unlikely that the same person had painted both.

Mindful of ramifications which could run into millions of dollars — and with a nod to novelist Dan Brown's bestselling thriller *The Da Vinci Code* — *The*

Marek Edelman

NAJWAŻNIEJSZE JEST ŻYCIE,
A KIEDY JUŻ JEST ŻYCIE,
NAJWAŻNIEJSZA JEST WOLNOŚĆ.
A POTEM ODDAJE SIĘ ŻYCIE
ZA WOLNOŚĆ...

Guardian headlined its article on the art scandal 'The Daffodil Code'. Questions remained unanswered. If da Vinci did not paint the National Gallery's image, who did? Why would anybody attempting to copy da Vinci's style invent as fantastical a flower as the chimeric *Narcissus tazetta*? If da Vinci did create it, what did that daffodil really represent?

The notion of meaning in art is one thing. In science it is quite another. In 2002 British biotechnology expert Professor Trevor Walker started to wonder if *Narcissus tazetta*, or one of its relatives, might represent hope. Walker had heard that a friend's wife had been diagnosed with the chronic, progressively neurodegenerative Alzheimer's disease at the age of fifty-eight. He knew of a naturally occurring alkaloid called galantamine that seemed promising in slowing the onset of the early stages of the disease. Galantamine had been isolated from the *Galanthus woronowii* snowdrop in the USSR, and Walker, who had spent time in Eastern Europe, understood that galantamine had been experimented with as a potential treatment for neurological conditions. He also knew that drug companies struggled to find inexpensive supplies of it.

Some daffodils contain galantamine, leading Walker to wonder whether their size might make them easier to work with than tiny snowdrops. Given that the United Kingdom is the largest producer of daffodil cut flowers in the world and generates around half of all narcissus bulbs sold annually, he speculated that local farmers might be able to grow daffodils medicinally, as a galantamine crop.

Narcissus tazetta contains this alkaloid but daffodils with larger bulbs appeared to have more potential. 'Carlton', a tall, golden, large-cupped Division 2 flower registered by Percival D. Williams with the Royal Horticultural Society in 1927, possessed good quantities of galantamine within large bulbs measuring 4 to

5 centimetres in diameter. Trials began with 'Carlton' and other potentially high-yielding daffodils to work out the optimum conditions in which to grow them (mild, wet and at altitude — making the Brecon Beacons region of Wales ideal) and when to harvest.

News of the quest reached Brecon Beacons sheep farmer Kevin Stephens, who in 2008 offered to grow daffodils for Walker's research and provide support for the fledgling business. Stephens has been involved ever since, in what he describes as an arduous journey. He told me of the first galantamine-processing machine they constructed: 'I built [it] from an old deep freeze, some bits from a milking parlour, a push bike and an electric drill.'

By 2011 Stephens had built a factory and established a novel way of producing galantamine by the kilo. The following year Stephens became director of a new company, Agroceutical Products, to buy up the assets of Walker's first business, Alzeim, which had gone into liquidation. The farmer-turned-entrepreneur spins a dramatic tale of funding challenges and developmental hurdles as Agroceutical has finessed methods of harvesting, extracting and storing the crop obtained from carefully selected daffodil strains. Stephens says he hopes to begin selling galantamine in 2016.

Alzheimer's is the most common cause of cognitive decline in older people, affecting up to 44 million worldwide at an estimated global cost of US$605 billion. To date neither a cure nor preventative treatment exists, making the spectre of a galantamine-based regime to slow the disease's progress a welcome possibility indeed.

But galantamine is one of many *Narcissus*-generated possible health-restoring alkaloids. The daffodil genus contains more than 300 alkaloids and, speaking as somebody touched by cancer, a number of these strike much closer to home.

Dig into the history and it is clear that the daffodil has a long association with cancer. Herbalists in the Middle Ages from China to North Africa co-opted narcissus oil in their attempts to combat it and earlier still in around 400 BC Hippocrates, the fabled physician of Ancient Greece, advocated combating 'female tumours' with narcissus flower ointment. Hippocrates, at least, may have been onto something.

Researchers have established that lycorine, the first alkaloid identified from *Narcissus pseudonarcissus* in 1877, shows promise in inhibiting ovarian cell cancer growth, as does narciclasine in the treatment of primary brain cancers. Jonquilline, a newly identified pretazettine alkaloid, has demonstrated anti-proliferative effects against a raft of cancers including glioblastoma, melanoma, uterine sarcoma and drug resistant lung cancer cells.

Yet *Narcissus* has a dark side. One reason gardeners like daffodils so much is because they are toxic and the wildlife knows it. Squirrels and muntjac deer that are forever nibbling the shoots off my mother's rose bushes never go anywhere near her daffodils. Human beings do not always demonstrate such good judgment. In 2012 ten Bristol residents accidentally bought pre-bloom cut daffodils in a supermarket thinking they were Chinese chives, cooked them up, ate them and suffered very serious episodes of vomiting.

Concerned that episodes of mistaken identity would continue, the government body Public Health England wrote to major retailers urging them to keep *Narcissus* bulbs away from food display areas in a bid to curtail inadvertent daffodil poisoning, of which there had been sixty-three cases in the previous six years.

Daffodil toxicity is a serious business. This plant contains poisonous alkaloids in its leaves, stems, bulbs and seedpods, and dogs have died after eating it, as did some Dutch cattle during World War II when their owners ran out of food and accidentally fed them a fatally poisonous dose. Humans who eat daffodils typically

recover after an unpleasant bout of nausea and diarrhoea, but those handling cut *Narcissus* also have to be wary of 'daffodil pickers' rash', a form of dermatitis caused by the combination of alkaloids and razor-sharp calcium oxalate crystals that reside within the plant's sap.

Daffodils can also be lethal to other flowers, a phenomenon called the 'vase effect', which scientists have used narcissi, tulips and roses to study. One experiment involved putting ten cut stems of 'Carlton' daffodils in a vase with ten cut red roses. After four hours at 2 degrees Celsius the rose flowers were visibly dying, their petals fading to a bluish red and their leaves deteriorating.

Other tests established that even short exposure to cut daffodils in a vase could prevent iris, freesia and anemone flowers from opening, while exposing iris flowers to a 'Carlton' stem provoked a remarkable delay in deterioration. Flowers would eventually wilt yet retain their original colour. Researchers surmised that the toxic alkaloid narciclasine seeped from daffodil stems into the water and prevented degradation.

Over the course of human history, healers have experimented with the notion that from poison can come cure. Ancient medics attempted to use daffodils as everything from an antispasmodic, emetic, analgesic, cure for baldness, aphrodisiac and contraceptive; suggesting it might remedy conditions as varied as epilepsy, bronchitis, whooping cough and even freckles. The Roman writer Aulus Cornelius Celsus advocated daffodil roots and seeds as a caustic 'erodent' to clean wounds in the first century BC, a practice echoed in traditional Japanese medicine where wounds were once dressed with a concoction of *Narcissus* root and flour paste.

To Nicholas Culpeper, the seventeenth-century English botanist and physician, yellow daffodils stood 'under the dominion of Mars', possessing roots which 'are hot and dry in the third degree'. He recommended boiling them up in a brew supposed to arrest spring fevers, and considered the 'White Daffodil' from France useful in curing (when mixed with vinegar) spots and noxious ulcers, as well as being handy for drawing splinters out from beneath the skin if blended into a honey poultice.

There were times when chemotherapy felt to me as ferocious as these brutal and seemingly absurd remedies. The crucial difference, of course, is that medical solutions of the 21st century have the weight of science behind them, and as I journeyed through the landscape of cancer treatments I realised that a large part of the daffodil's true power actually lies elsewhere.

In 1920s England, wild daffodils became an annual gift from the country to the city's hospitals. Silent black and white Pathé footage from 1926 entitled 'Daffodil Day' shows children crowding through the Gloucestershire market town of Newent carrying thick bunches of daffodil blooms.

The flowers travelled on the so-called Daffodil Line, the Ledbury and Gloucester Railway that each spring would transport tightly packed crates of narcissi destined for sale in the Midlands and London. The British Legion oversaw the delivery of 20,000 daffodil bunches to forty different London institutions in 1929 alone.

Around this time in Canada the medical community was trying to come to grips with cancer. Doctors had two issues to deal with — blunt-force treatments and under-educated patients, too many of whom consulted medical specialists only when the disease was too advanced to treat effectively. By 1938 the Canadian Society for the Control of Cancer (later renamed the Canadian Cancer Society) had been formed, and those agitating for change realised that fundraising was key.

In the early 1950s a Society volunteer called Jack Brokie, who worked at the prestigious Toronto department store Eaton's, convinced Lady Flora Eaton, the owner's wife, to preside over a high society springtime fundraising event called

a 'Daffodil Tea'. Organisers decorated the tables with bright narcissi and 700 women attended.

During the spring of 1956, to mark the Society's first day of fundraising, Toronto restaurateurs agreed to donate some of their proceeds to the cause. Volunteers offered generous patrons daffodils to say thank you, prompting some customers to offer even more. It prompted a brainwave. Was it possible, when it came to daffodils, that people would willingly pay for them?

In 1957 the organisation tried just that on its first ever 'Daffodil Day' during which volunteers roamed the streets of Toronto selling the flowers to the public — an experiment that raised the sizeable sum of $1,200. The daffodil became the Canadian Cancer Society's official fundraising emblem, and though the Society's records hold no clue as to how much live flowers have raised since then, in 2013 alone the sale of 5.5 million fresh daffodil flowers generated over C$4.7 million. Health organisations in other countries took notice of the Canadian initiative, and the daffodil continues to be used to raise funds for cancer research all over the world.

CHAPTER 8

The Fifth Element

'Then comes the question of
the corona or trumpet.'

F.W. Burbidge,
The Narcissus: Its History and Culture, 1875

Moments of true revelation in science are astonishingly rare and the last place Oxford University biologist Dr Robert Scotland ever expected to have one was on a Portuguese cliff top, during a routine undergraduate field trip while peering into the heart of a tiny wild daffodil.

Scotland remembers the instant precisely. It was a clear spring day in 2004 in the Algarve, the most southerly area of mainland Portugal overlooking the Atlantic Ocean. Every spring a champagne-tinged carpet of miniature native daffodils, the delicately fluted *Narcissus bulbocodium*, adorns this coastal headland. The Royal Horticultural Society classifies it as a Division 13 daffodil, one distinguished solely by its botanical name, but this delicate miniature was known in Peter Barr's day as the Hoop Petticoat Daffodil because its flower shape reminded people of the undergarments women used to wear. For the Oxford biologist both the daffodil and the landscape represented very familiar terrain.

Scotland had spent years bringing students here to show them how plants adapt to and thrive in the Mediterranean climate's long dry summers and cool winters. On this occasion, as usual, he spent the first few days making sure the students had a thorough grasp of floral architecture.

'Even though these are second year undergraduates they do need introducing to basic flower anatomy', he told me, explaining as he would to his class: 'There are only four basic parts to flowers — the petals, sepals, stamens and carpels — but they are modified in various ways.

'The sepals are often green, a photosynthetic structure; they protect the flower and the bud. Inside the sepals are the showy petals. With some spring-bulb plants, like daffodils, the sepals and petals look the same; the same colour, same shape, same size. For such flowers we use the term "tepal".'

On this particular day in the Algarve, Scotland ran through all this with his students, as usual. He told them to examine one of the little native daffodils and primed them on what to expect — three sepals on the outer edge of the daffodil and three petals, looking very similar, within. Each pupil methodically counted their way through the four flower parts of their daffodil: the outer and the inner two whorls, then within, the stamens (the male parts), and the carpels (the female parts) in the centre. In part, says the botanist, this exercise demonstrates that while there are some spectacular exceptions the vast majority of the 350,000 species of flowering plants in existence, including these dwarf daffodils, fit this basic pattern of floral architecture. There would be no surprises here.

Except that one of Scotland's students did not quite understand. From what she could see her daffodil contained something else, an element that did not fit into her tutor's four-part structure. 'So what part of the daffodil is the trumpet?' she wanted to know.

'Ah, that's an interesting question,' he parried, directing her to go through what he was teaching one more time, step by step. 'Work from the outside to the inside, identify the sepals, the petals, then the stamen, then the carpels,' he told her watching as she methodically went through the process once again, he suddenly realised, 'If you do that with the daffodil you get all those parts and you're still left with this large trumpet-like structure coming out of the centre. At which point I realised that I could not answer her question.'

With a hollow laugh Scotland admits he has never experienced such an instant before or since. Having been raised in the United Kingdom he had, quite literally, grown up with daffodils. He must have seen thousands over his lifetime. More than that, as a botanist, he had studied them repeatedly on field trips like this. Yet he suddenly realised that he had never looked at them properly. He had absolutely no idea how this flower worked.

'The daffodil is such an iconic structure, it is so well known ... and it's not as if there are a hundred different ways to make a flower,' he told me. 'I'm a tropical botanist and people tend these days to think, "Oh the mystery is in tropical rainforests." Daffodils are common European garden plants, and I did not know. It was a very sobering moment.'

When Scotland got back to Oxford University he contacted two world-class plant anatomists he felt sure would easily be able to settle the matter, one from the Jodrell Laboratory at Kew Gardens, the other from Zurich University. He was wrong again. Neither could help.

They directed him instead to a debate that had begun in the nineteenth century between those who argued that the trumpet, or 'corona' as it is termed, was a severe modification of the petals, and those who believed that it had to be an extreme adaption of the stamens. Yet nobody knew for sure. The daffodil had stumped them all.

'Sometimes the things that seem very obvious are the things we know least about,' observed Scotland, who decided to resolve the mystery himself by launching an experiment that would run for eight years as a side project to his primary work. He secured a grant of £10,000 and began researching in his spare time.

'One of the difficulties, something I didn't know about daffodils then, was that no one had been able to look in a really detailed way at the early stages of how a daffodil flower develops,' he says. A daffodil will flower gloriously in spring and then fade away, withering back down to the ground. As the summer months progress the following year's flowers are created within the bulb, in microscopic form. Scotland theorised that he should be able to find these minuscule structures within the *Narcissus* bulbs — 'absolutely tiny but fully formed', as he put it — even though they would measure probably just a couple of millimetres in diameter.

Therein lay the rub. To solve the mystery of the daffodil's trumpet he would have to discover precisely how each separate structure within the flower is made. To do that the scientist had to be able to observe those earliest stages of development as they took place within the bulb, which meant, in the first instance, buying up an awful lot of bulbs.

Scotland decided that because the Portuguese Hoop Petticoat Daffodil had sparked the question, that variety should be the one to sort it out — a choice he would come to rue. 'They have small bulbs and flowers,' he said. 'In retrospect we should have picked a different one. A larger one.'

In the years that followed Scotland's team of students dissected thousands of bulbs, blindly hoping to find embryonic flowers inside. 'We would cut into them but the easiest way to do it was just to peel each one by hand, and I could only do that for about ten days before my fingers were so sore and bloody that I would have to stop,' he said of a process that turned out to be 'completely hit and miss'.

Even when they did get the timing right and find a miniature daffodil flower within the miniature bulb, 'our dissecting needles would go through the centre of them because they were absolutely tiny'.

When they did finally manage to isolated the flower without damaging it, the researchers had to carefully remove each one for examination and detailed dissection. It took several more years before the team managed to take the first, landmark electron-micrographs — of images nobody had managed to photograph before.

The first thing Scotland wanted to understand was whether the Hoop Petticoat Daffodil's corona developed as a circular structure, or grew into one. The answer appeared to be that it initiated in six different places between the petals and the stamens. Those tiny sections of tissue then joined together to form the corona's ring.

But this only explained so much. Exactly what *was* the corona? Answering this involved identifying and tracking the daffodil's 'master genes', in particular what is known as its 'B gene', which is active in petal formation, and its 'C gene', which works in tandem with the 'B gene' to form the stamen.

'What we were able to show is that the corona is made from B *and* C genes, just like the stamen, even though it is an independent organ,' says the scientist. 'That was detailed, fiddly, molecular biology because we had to clone the genes and show whether the genes were being expressed in the tissue of these very tiny, developing flowers. It was technically quite challenging.'

It solved the mystery, though: the daffodil's corona is neither part of the stamen or the petals. It represents a fifth element — something quite new in flower architecture.

The resulting scientific paper's authors, led by Scotland, included Oxford University geneticist and plant scientist Professor Jane Langdale, United States Department of Agriculture botanist Alan W. Meerow, Bussey Professor of Organismic and Evolutionary Biology at Harvard Elena M. Kramer, and University of Western Australia biologist Dr Mark Waters. The paper ran in the prestigious scientific journal *Nature*; no small achievement for a low-budget, part-time project first fired by a single question and pure curiosity. The enigma had pitted the latest tools of molecular biology against an ancient anatomical debate: 'A mixture of old questions and new techniques', as the researcher put it. It also, at times, threatened to overshadow his other work.

Upon being interviewed by an Australian radio show on the topic of his findings about plant cataloguing — 'We had found out it takes thirty-five years between when a plant is collected to when it's described as a new species, that is a huge delay' — Scotland was asked at the end of the prerecorded conversation whether he had been working on anything else.

'I said, "Yes, I've been trying to work out what the trumpet of the daffodil is," and that ended up being the entire radio broadcast,' he relates with a laugh. 'They cut the rest.'

Asked if he looks at daffodils differently now, the botanist replies with frankness, 'Yes, I do. I think of my nails bleeding from peeling thousands of daffodil bulbs. It reminds me of the labour of science, the pain of that always comes to mind.' Even so, he adds quickly, 'I'm always pleased to see a daffodil.'

Scotland's detective work solved a centuries-old mystery, yet another riddle lingers — the issue of what the corona is for. 'That's not a question I've really been thinking about,' he told me. 'I don't think anyone really knows the answer to that.'

Evolutionary biologist Professor Spencer Barrett from the University of Toronto, might disagree. Barrett spent a decade studying the sex lives of daffodils and notes of Scotland's research project: 'You would not have been able to do [it] even twenty years ago because molecular genetics and developmental biology were nowhere near as sophisticated as they are today.'

Barrett's own interest in *Narcissus* was, in a sense, equally opportunistic; a question of 'Now is the right time because there is this new suite of approaches.' Yet as Barrett expounds on his own intense relationship with daffodils, it becomes clear that he sees himself rather as the daffodilian amateur hybridisers do, as

someone who inhabits a landscape where many have walked before. He wanted to tell me a daffodil story of his own, one that involved 'British intellectual chauvinism and changing viewpoints', and it began with the man who changed science forever.

In 1842 Charles Darwin moved his young family from the cramped confines of London into Down House, a three-storey Georgian home in the quiet county of Kent. Down House had everything the ambitious, thirty-three-year-old naturalist needed: an estate with gardens, meadows, an orchard and plenty of greenhouse space.

Here Darwin crystallised his revolutionary theories. He planted, grew and experimented with the botanical plant specimens he had brought home from his travels, coalescing his thoughts in *On the Origin of Species by Means of Natural Selection*, one of the most significant books ever released. To this day, for many, Darwin's legacy remains dominated by his groundbreaking work on the evolution of life; yet he was profoundly fascinated by the baffling realm of flowering plants, a topic he would publish on repeatedly and spend longer studying than he would anything else.

On the Origin of Species reflects both his interest and the reason for it. In it Darwin discusses mistletoe, the flowers of which have separate sexes, but he muses about blooms that are not only hermaphrodite but 'in which anthers and pistil ... stand so close together that self-fertilisation seems almost inevitable'.

'How strange that the pollen and stigmatic surface of the same flower, though placed so close together, as if for the very purpose of self-fertilisation, should in so many cases be mutually useless to each other!' he wrote. 'As yet I have not found a single case of a terrestrial animal which fertilises itself.'

Darwin had realised the vast majority of flowering plants are not only hermaphrodite but have male and female organs positioned closely together within their flowers. So, can they fertilise themselves, and if they do what will their offspring be like?

'Darwin became obsessed with this,' says Barrett. 'First he tried to prove that maybe they weren't affected by inbreeding. He worked for ten years "self-ing" and "crossing" around fifty-odd [plant] species only to find in every case that yes, they were affected.'

Darwin's inbred plants not only grew less vigorously than those with unrelated parents but tended to be sterile, explains Barrett. 'The conundrum was sealed. They are affected by inbreeding but they are hermaphrodites. So how do they stop it?'

In humans incest is a cultural taboo and for Charles Darwin the issue was personal. In 1839 he married Emma Wedgwood, his first cousin on his mother's side. Compounding the issue, Darwin's maternal grandparents were third cousins. Of Charles and Emma's ten children three died young and another three, despite being in lengthy marriages, would produce no offspring of their own.

Tim Berra, Professor of Evolution, Ecology and Organismal Biology at Ohio State University, published his findings on the convoluted Darwin family tree in 2010, suggesting that Darwin's concerns about the dangers of inbreeding were well founded. 'He fretted that the ill health of his children might be due to the nature of the marriage and he came to that because of his work on plants.'

Darwin's experiments yielded several major botanical discoveries about how plants sidestep incest. Being incompatible with themselves is one method, creating flowers of different types — a phenomenon called 'heterostyly' — is another. Plants that use this tactic possess flowering forms in which the style, a long slender stalk, connects the stigma (where the pollen is deposited) and the ovary. Distylous plants have two of these flowering forms: long and short. Tristylous flowering plants have three: long-style, mid-style and short-style.'

The trait appeared to have evolved independently in many different flowering

plants, and what Darwin wanted to know was, 'Why?' According to Barrett, that answer ricochets back to inbreeding because it is a mechanism to stop self-pollination between flowers. 'The long-styled form can only mate with the short-styled form and the short-styled form can only mate with the long-styled form,' Barrett explains. 'It's an odd version of hermaphroditic sex where there are either two mating types or three mating types.'

Darwin died in 1882, five years before Portuguese botanist Júlio Augusto Henriques in 1887 and 1888 described two species of daffodil as tristylous plants. Around forty years later Abílio Fernandes, another Portuguese botanist, not only confirmed Henriques's findings but counted the frequency of occurrence of the three different tristylous flower forms, and speculated on the genetics behind them.

In the early 1950s Angus John Bateman, an eminent scientist Barrett describes as one of the 'British genetics mafia' who read Fernandes's work. He wrote to the Portuguese researcher asking for daffodil bulbs. Fernandes complied and Bateman planted, grew and crossed the plants, then examined the flowers that resulted. Bateman concluded, in a 1952 *Nature* paper, that Fernandes was wrong.

Bateman's views had weight. His work with fruit flies (*Drosophila melanogaster*) led him to hypothesise that males are inherently promiscuous, their reproductive success increasing with the number of mates they have, while females are fundamentally selective when it comes to sex. In the annals of science researchers are seldom immortalised by having theories named after them but this one is known as 'Bateman's principle'.

Spencer Barrett learned about Bateman and his assessment of Fernandes's work as he studied heterostyly, which he describes as his own life's work. Unconvinced, he wondered if Bateman's findings had more to do with intellectual snobbery about this quintessentially 'British flower' than good science.

'What does this Portuguese guy know about daffodils?' Barrett asked rhetorically, as he suspects Bateman might have done. 'Well, he should know something about daffodils because they are native to Portugal, and it's a disputed issue as to whether they are truly native to the British Isles at all.'

So Barrett began reviewing those early Portuguese research papers. At the time he happened to be in contact with David Lloyd, a New Zealand evolutionary biologist, and the pair agreed to study the original daffodils in the wild. Fernandes was still alive, so they visited him in Coimbra, Portugal. He showed them where on the Iberian Peninsula to find the daffodils, and so began what Barrett depicts as a 'wonderful decade' where he would escape the bone-chilling Toronto springs to hunt narcissi across Portugal, Spain and Morocco.

Barrett and Lloyd embarked on their first joint field trip in the early 1990s. They flew to Seville — Barrett from Toronto and Lloyd from Christchurch, New Zealand — rented a car, and came across an extraordinary oak woodland full of daffodils. 'We sat there speechless,' the Canadian scientist recalls. 'He sat cross-legged in one section of the population while I sat in another, down in the grass with them.' Barrett watched his colleague lying in among the daffodils so he could study exactly how the bees bumbled into the flowers in search of pollen.

'Just being with David trying to figure out ... how do these flowers work?' says Barrett, attempting to define the subtle qualities of the perfume they encountered when they entered fields awash with Paperwhite Daffodils and what it was like watching the long-tongued hawk moths the flowers attracted. 'Waiting for pollinators to come in; seeing how do they enter, where do they get pollen on their body, how is cross-pollination being effected here?' Those days were, Barrett sighs, 'Fantastic.'

Lloyd was captivated by how daffodils wield their power, in particular the question of how these flowers can control their sex lives by manipulating the insects that disperse their pollen. Studying these daffodils would lead Lloyd 'to several novel insights that had eluded generations of pollination biologists,' says his colleague.

They discovered that Fernandes had been quite right about the daffodils. 'He died before we published ... but he was completely unfazed by it,' Barrett says. 'I had asked him: "What do you think of the Bateman criticism?" and he just told me: "He is wrong."'

Barrett decided other fundamental questions about daffodils needed to be answered and tackled the sticky issue of how different species were related. He and Sean Graham, now a professor at the University of British Columbia, collected *Narcissus* DNA and, using molecular biology techniques, established a family tree.

Yet Darwin's puzzle lingered: the daffodil is hermaphrodite so what botanical chastity belt stops it fertilising itself? Try to force it to by manually putting a daffodil's pollen on its own stigma and no seed will result, Barrett found. The daffodil foxed every attempt.

Flowering plants generally use one of two techniques to block self-fertilisation, he advises. Either the stigma (the female tissue) refuses to let the pollen grain's pollen tube grow in it, or the pollen tube grows through the stigma towards the ovary only to be rejected by the plant halfway down the style.

Which method did the daffodil opt for, Barrett wondered. His team's subsequent investigation revealed the answer was neither. To their surprise they realised the pollen tubes travelled all the way down the style and into the ovary — exactly what happens when cross-pollination takes place — but as the pollen tube enters the ovary the ovule collapses.

'The maternal plant just shuts its female function down,' Barrett states bluntly. 'That is contraception.' To the best of his knowledge this phenomenon is rarely seen in the plant world. 'It may have subsequently been discovered in one or two other species but our observation was the first.'

In answer to the question of what the daffodil's corona is for, he replies, 'My interpretation ... is that coronas are a mechanism that positions bees in such a way that they pick up pollen on their body and effect in an efficient way cross-pollination.' Darwin was quite right, he adds, 'It is back to inbreeding.'

According to the Royal Horticultural Society database there was once a daffodil named after Charles Darwin. Bred by Henry Backhouse (1849–1936), one of the sons of William Backhouse, it first flowered before 1908 and is listed as a Division 10 bulbocodium cultivar, meaning that it is genetically related to those elfin Hoop Petticoat Daffodils that carpet the coastal headlands of Portugal each spring.

CHAPTER 9

Scents and Sensibilities

'Styles change! The length of skirts
or hair, the amount of decolletage
or gaping seam are matters of
continuing judgment and consideration.
Styles in daffodils change, also!'

Dr Tom D. Throckmorton,
'Maybe It's Time for Recess' *The Daffodil Journal*, December 1979

Narcissus poeticus is an antique daffodil with an otherworldly beauty. It blooms late in spring, is believed by some to be the variety alluded to in ancient Greek legend, and displays a simple halo of pure white arching petals and a startlingly yellow scarlet-rimmed crown. My mother, like so many others, thinks of the *Narcissus poeticus* var. *recurvus* variety that grows in her garden as 'Pheasant's Eye', and while *poeticus* may not be the daffodil that so inspired William Wordsworth it is imbued with a strange lyrical power.

Sir Bernard Burke, the nineteenth-century genealogist who created *Burke's Peerage*, realised this while investigating the enigma of the fallen house of Finderne for his 1860 book *A Second Series of Vicissitudes of Families*. Once a rich and powerful Derbyshire clan, the Findernes mysteriously vanished in the mists of the fifteenth century. The village, Finderne (now spelled 'Findern') still exists but when Burke explored it he could find no trace of the Finderne dynasty, so he accosted an elderly local for information.

'Findernes,' repeated the old man, 'We have no Findernes here but we have something that once belonged to them.' He led Burke to a field containing traces of ancient ruins, and pointed to a bank of wild blooms. 'They are the Findernes' flowers, brought by Sir Geoffrey from the Holy Land,' said the villager of the aristocrat's campaign in the Crusades around eight centuries earlier, 'and do what we will, they will never die.'

These flowers are believed to have been *Narcissus poeticus* and their presence deeply affected the author. 'For more than three hundred years the Findernes had been extinct, the mansion they dwelt in had crumbled into dust, the brass and marble intended to perpetuate the name had passed away,' Burke wrote.

'A tiny flower had for ages preserved a name and a memory which the elaborate works of man's hands had failed to rescue from oblivion. The moral of the incident is as beautiful as the poetry. We often talk of "the language of flowers" but of the eloquence of flowers were never had such a striking example.'

Narcissus poeticus cast its spell over some of the daffodil's most impressive experimenters. Margaret 'Meg' Yerger (1914–2008), whose achievements included becoming one of America's earliest female pilots, pioneered the hybridising of miniature poeticus and registered around 100 new varieties. Alexander M. Wilson ('father' of around seventy named poeticus cultivars), P.D. Williams and Guy L. Wilson also became enchanted by the breed, as was George Engleheart, who developed a habit of giving his most splendid 'poets' the names of actual poets such as 'Chaucer', 'Dante', 'Horace' (one of this breeder's greatest daffodil triumphs), 'Tennyson' and 'Virgil'.

The lure of *Narcissus poeticus* lies within the intense contrast between its flower's seemingly innocent fragility and its unsettlingly erotic smell. One of the few *Narcissus* species used in fragrance creation (*Narcissus tazetta*, *Narcissus odorus* and *Narcissus jonquilla* are other key varieties), it is by far the most intriguing, and from it comes an essence that is distinctive, precious and rare. Perfumers are masters at attempting to define the indefinable, yet they find themselves pushed to their limits of sensory conviction with their descriptions of this flower.

Narcissus poeticus is a major challenge. Its heady aroma is complex, a melange of around 300 different chemical components, and the terms used to characterise it play with entangled ideas that are at once dark and light, delicious and disturbing, mysterious and animalistic, elusive and direct. Even the floral notes appear, at first glance, contradictory — meshing orange flowers, jasmine, violet and rose.

Fragrance creation is a realm where art, science and commerce intersect with the deeply personal. British artisanal practitioner John Bailey, a former president of the British Society of Perfumers, considers *Narcissus* a potent raw material to work with and says it can be found, unmentioned, in many high quality perfumes. He describes its pungent scent as 'narcotic, almost intoxicating' and reaches for it, and the scent of jonquil, when conjuring an aroma that requires 'velvety body and depth'.

The most concentrated fragrance form of *Narcissus poeticus* is called 'narcisse absolute'. It is supplied commercially to the global scent industry by International Flavors & Fragrances (IFF), whose publicity material uses the words 'green', 'intoxicating', 'rich' and 'mellow' to define it, explaining that it possesses a 'floral character' with 'hay and honey undertones'. Unlike some of the other florals this firm produces — such as lavender which is inexpensive and used in everything from shower gel to fabric softener — narcisse absolute is exclusive

PL.XLII

F.W Burbidge del et lith.

V. Brooks Day & Son Imp

N. POETICUS.

A .... RECURVUS.

and elusive; destined for high-end companies such as Cartier and Hermès that focus on luxury goods.

IFF's Laboratoire Monique Rémy research centre can be found in a nondescript industrial park a short drive from the charming, medieval southern French city of Grasse, Europe's history-laden perfume capital. Accessible by appointment only, its doors open to reveal walls sporting large colour photographs of *Narcissus poeticus* flowers and other raw ingredients, and white-coated scientists busy perfecting methods of transforming plant material into powerful fragrance forms.

This is an intense, alchemic, aroma-laced environment. Machine rooms hold metallic-and-glass devices of baffling appearance which thunder away as they crush, pulverise, distil and refine. Bulbous test tubes of varying sizes glimmer with concentrated distillates as mechanised mixing devices stir what appears to be heavy, golden dough. Every now and again a machine blasts dry ice into the atmosphere.

Despite the hallmarks of wizardry this place is state of the art, and scientific to its core. The processes underway are tested and true — be they steam and molecular distillation, CO_2 extraction or fractionation, these are technologies proven to create bespoke, high-quality results. Research and Development Manager Jean-Claude Bayle walks me through the complex past a small, secure storeroom where millions of euros worth of extracts are held. Narcisse absolute is one of the scarcer and most highly priced.

For the flower itself the story begins 500 kilometres away in the Massif Central, a mountain range where, as coincidence would have it, my sister happens to live. Here on the windswept Aubrac highlands 1,300 metres above sea level in a remote, barely populated district of France called the northern Lozère, *Narcissus poeticus* blossoms each year.

For decades the farmers who work this harsh terrain have supplemented their hard-fought income by selling the wild daffodils to perfumers, but harvesting is a challenge. In the early days these farmers plucked the delicate flowers by hand, being careful not to damage them. Then they developed a rudimentary, wooden rake that gently combed the blooms away from their stems. Adding wheels to the wooden combs sped up the process and enabled farmers to carefully harvest up to 100 kilograms of daffodil heads per day, but the task remained laborious and time-consuming.

To assist them, and ensure a steady supply for itself, IFF helped to develop a machine nimble enough to deal with *Narcissus poeticus* flowers yet capable of coping with the formidable terrain. It can reap up to one tonne of flower heads daily. *Narcissus poeticus* has a two to three week season, usually in May, and farmers have to know exactly when the blooms start appearing. 'They flower for one day,' says Jean-Claude Bayle of the plants, and the need to pick at the perfect moment. 'It is necessary to crop rapidly.'

The cut flowers are rushed to warehouses and strewn, dense and ankle-deep, across the expansive floors. They are turned continually by workers wielding large forks, and then sealed inside solvent extracting machines that capture their aroma, in the form of a concentrate, before their scent degrades. As soon as the harvest begins the work goes on day and night to swiftly process the fragile flowers. A good season generates a crop of 100 to 150 tonnes of flowers, which are transformed into two forms of perfume essence — Narcisse Absolute French, which has a shelf life of twelve months, and the long-lasting Narcisse Absolute French CO_2. It takes about 1,600 kilograms of *Narcissus poeticus* flower heads — approximately 800,000 individual flowers — to make 1 kilogram of the former, and 3,200 kilograms to create 1 kilogram of the latter. This daffodil is harvested for perfumery in areas of Morocco and Egypt as well as this isolated region of France, but according to Judith Gross, IFF's global marketing director, narcisse absolute would not exist were it not for her company.

'These flowers are scarce and without the new harvesters it would not have

been worthwhile for these farmers to crop it,' she elucidates, adding that the supply would have shrunk, forcing the already high price to 'skyrocket'. 'There is a tipping point and when you reach a certain price the perfumers stop using it,' Gross explains. 'They won't use less; they just stop. So if it had not been for that work we conducted with the farmers there would be no more narcisse for perfumery at all.'

My sister knows of these mountain farmers but as producers of Aubrac cattle: a hardy, muscular, tan-toned breed selectively reared by French monks hundreds of years ago. How long *Narcissus poeticus* has inhabited these inhospitable uplands remains a mystery but my sister sees its flowers blush across the savage Gallic landscape every spring. My mother witnesses the same thing over a thousand kilometres away in her comparatively tame corner of southern England, and being what you might call a casual gardener, *Narcissus poeticus* is perfect for her as it basically thrives on being ignored. She once reacted to its appearance by cutting some blooms and taking them to a nearby village spring show, where they delighted her by demonstrating an aptitude for winning a first in the 'Poeticus' category.

No poeticus that I am aware of has ever crash-landed through the media in the way Division 2 'classical daffodils' are able to, yet of all the varieties that exist it is the one that means the most to me.

That fact began to dawn on me, ever so gradually, as my treatments for cancer progressed. My therapies had side-effects and as the months went on I grew progressively weaker. The regime involved a series of chemotherapy sessions at my hospital's oncology unit, each of which left me fighting dizziness, nausea and dread. I also developed what is colloquially termed 'chemo brain', a curious byproduct of the chemical cocktails that many manage to avoid completely. It manifested as a mental fog that left me with barely enough concentration to read and the terrifying fear that I might never again be lucid enough to write.

Through the brightest of Sydney summers, in city heat that topped 40 degrees

Celsius, I felt trapped in an endless winter night. I repeated a mantra, that things would get better, that they had to get better — and with that thought to my surprise came the sharp image of *Narcissus poeticus.*

I have known this flower all my life and suspect I have always loved it — for its individuality, its softly luminous 'petals' and that irrational corona that looks for all the world as though somebody decided to dip it in a pot of bright red ink. My brother sent me photographs of many of my mother's different daffodils but it is his images of *Narcissus poeticus* that I kept coming back to. They captivated and nourished me.

As my treatment ended, I grew stronger. I journeyed from Sydney to visit my parents without thinking too much about the daffodils. It was so late in the season that tight-knit clumps of bluebells had already begun emerging, in fits and starts, across the nearby woodland floor. I went looking for daffodils even though I suspected it was far too late to find them. The ground was still covered in thickly packed *Narcissus* leaves but the only flowers visible were dead, diseased or dying; what was left of them shrivelling back grotesquely into a swaying sea of green.

Yet one by one, as if from nowhere, a few late-flowering daffodils began to appear. First 'Mrs R.O. Backhouse', so elegant with its pink cup and flowing white perianth; then 'Sulphur Phoenix', an ancient, frilly and really rather silly double daffodil that froths with cream and orange ruffles, and is known to my mother by the nickname 'Butter and Eggs'. Another peculiar, all-white ghost of a *Narcissus* nodded as though beckoning me to it. Attired in a crenelated petticoat of a corona, its slender, sweptback 'petals' indicated that it could be 'Thalia', a pre-1916 Division 5 triandrus bred by M. van Waveren & Sons in The Netherlands or — just possibly — the eerie 'Venetia', a daffodil registered by Henry Backhouse in 1910 that has not been seen for years.

I drifted further into the daffodil meadows and it was there that I saw it — a single, perfect *Narcissus poeticus* in bloom. Instinctively I did what I used to do as a tiny child. I dropped down onto my knees and then sank further until I was lying, front first, on the grassy earth. Around me *Narcissus* leaves and stems swayed to the breeze's silent rhythm. Insects darted into view and hovered. Time slowed down, then seemed to freeze. The little daffodil faced me.

When I remember that moment I can smell *Narcissus poeticus* still.

Daffy-down-dilly is now come to town
With a petticoat green and a bright yellow gown.

CHAPTER 10

Judgment Day

'All of us want to know what
the future holds.'

Jan de Graaff,
The American Daffodil Yearbook, 1960

Rupert Sanderson, a British advertising executive with dreams of becoming a designer, was trawling through a Sussex car boot sale in the spring of 2000 when a quirky book caught his eye. The slim green volume, *Classified List and International Register of Daffodil Names*, was printed in 1959 for the Royal Horticultural Society and catalogued thousands of *Narcissus* cultivars. 'I never knew there were so many: list after list, name after name,' Sanderson recalls. 'So for the princely sum of 15p I took it home.'

The following year as he prepared for the launch of his inaugural luxury footwear collection, he remembered the little title, decided it would become his muse, and christened each shoe style after a different daffodil. Sanderson, whose shoes now retail worldwide from high-end outlets, continues the tradition to this day.

'Our eye constantly shifts in terms of what we find appealing,' observes the designer. He finds the fact that hybridisers can contort daffodils into such different forms 'inspiring' and says when it comes to style and popularity, taste and fate weed out the weak. 'You have to keep moving forward with fashion or you're dead in the water,' he tells me. 'The same would be true with breeding daffodils.'

In 2011 Sanderson visited the Royal Horticultural Society's Great London Plant Show and met Adrian Scamp, who works with his father, Ron, in their Cornwall-based bulb business. Scamp junior showed Sanderson an exhibition area populated with avant-garde narcissi identified by number but, as yet, no formal name. It seemed like a 'daffodil waiting room' to the designer, who in 2013 decided to engineer a hybrid of his own. Using 1,500 Scamp-sourced blooms of 100 fresh narcissi he fashioned a 2-metre high stiletto shoe and, installing it in his stylish boutique in Mayfair, London, used it as a centrepiece for the celebrity launch of his opulent spring collection.

Two months later botanist Melanie Underwood happened to replace Sharon McDonald as Royal Horticultural Society's International Registrar for Narcissus & Delphinium, making her the official gatekeeper for the daffodils in Sanderson's imaginary 'waiting room'. The Society's database now holds the names of around 31,000 registered daffodils plus another 2,000 which are unregistered for various reasons, says Underwood. Many have been lost over the years, she adds, with only 'a small fraction' of the 31,000 known to still flower yet the list continuously swells.

Underwood receives around 200 requests a year from breeders around the world wanting to register novel narcissi. The most popular are the classic Division 2 large-cupped daffodils, of which around 12,000 registrations. Division 1 trumpet daffodils are next, followed by small-cupped Division 3s, and then the doubles from Division 4.

Each fresh daffodil must be properly listed with details of colour, division, raiser, year of cross, seed and pollen parent, and every new name should comply with guidelines set out in *The International Code of Nomenclature of Cultivated Plants*.

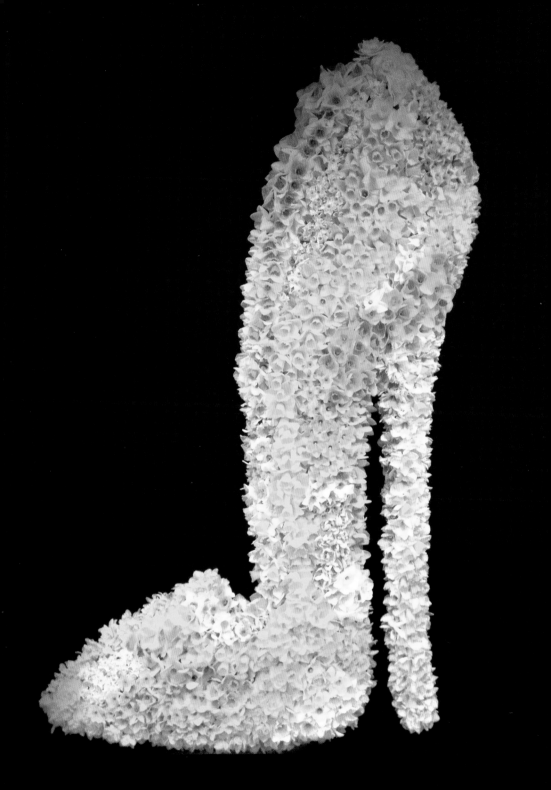

Fractions and symbols are banned, as are names confusingly close to those of existing daffodils.

'A cultivar whose epithet is or contains the name of a living person should not be published unless that person has given permission,' Underwood says. 'I had one recently for President Vladimir Putin and I can't imagine that permission has been given but I am waiting to hear back. It is just a recommendation, though. If people insist, we have to say, "OK".'

Some pick names with an eye to a headline, often to raise charity funds. British breeder Johnny Walkers, a serial Chelsea Flower Show medal-winner, made the national press by naming a white and yellow Division 2 daffodil 'Georgie Boy' after Prince George, the young heir to the throne. Ron Scamp has periodically registered daffodils after people for the same reason, such as his delicate white and pink Division 2 *Narcissus* 'Alex Jones' (named after the Welsh TV presenter) and another attractive Division 2 daffodil with an orange-red cup and butter-yellow 'petals' called 'Gallipoli Dawn', to commemorate the centenary of the 1915 Dardanelles campaign of World War I.

Of course beauty is in the eye of the beholder, and perfection a matter of judgment. New Zealand *Narcissus* grower and judge Henry Dyer (1898–1985) once arrived at a tiny country show to discover the amateur exhibits on display quite pathetic. One category forced him to choose between just two feeble daffodil specimens and as he agonised over which was the worst he noticed they stood in different bottles, one of which had once held beer, the other whiskey. With a shrug, he gave the whiskey first place.

The connoisseurship required for passing verdict over daffodils has been hotly contested by aficionados. 'It is with some misgiving and almost a sense of shame that I admit I am not an Accredited Daffodil Judge,' wrote Dr Tom D. Throckmorton in the December 1979 edition of *The Daffodil Journal*, during an editorial debate on the issue, wryly citing 'desolate geographical location' (he lived in Des Moines, Iowa) and 'lack of time' as his only excuses. Throckmorton was not just a hybridiser and former American Daffodil Society president but fathered the daffodil colour coding system still widely used in the *Narcissus* world. His argument that judging should take into account 'nuances of color, grace, symmetry, and all those other things which make daffodils both beautiful and individuals' seems difficult to take issue with.

'I cannot imagine a Miss America Beauty Contest being decided by a computer,' he added. 'Yet, I fear that some of our daffodil judges are accepting computer sources rather than the evidence of their own eyes and years of experience in daffodil cultivation.'

Jackie Petherbridge, president of the UK's Daffodil Society, has judged daffodils for over three decades and, like all accredited judges, abides by strict criteria, awarding points for well-defined characteristics. These include form and poise, how a flower 'holds itself', as she puts it, as well as the flower's condition, freshness and texture.

'The colour has to be correct,' she explains. 'Conversely if the daffodil is a reverse bicolour it has got to have faded so that the halo of white can be seen at the back. The flower's size has got to be right, because biggest isn't always best and the stem should be straight and true and proportionate to the size of the flower.'

Petherbridge has nightmares about being forced to choose a winner from thirty identical Division 2 white-cupped narcissi. 'The other thing is when you are invited to a village show where they give people bulbs and a pot and say go away, grow them and bring them back,' she says. 'You'll have ... pots of identical bulbs that people have brought in, their pride and joy, and it's up to you to say, "This is the top one", and then they all want to know why.'

Daffodil judges scrutinise thousands of blooms. They whittle them down to Best In Category, and then to Best In Show. Of that ultimate decision — and that all-important but nebulous quality — Petherbridge says, 'You pick the one that

captures you by its inner beauty. With a real winner there is just this extra aura. Perfection. Harmony. That's all.'

Petherbridge contracted 'yellow fever', as she calls it, around forty years ago during a talk by a *Narcissus* doyen who brought along ivory-petalled daffodils with whimsical pink trumpets. Petherbridge, who had never seen anything like them, fell instantly and enduringly in love.

'Daffodils never lose their spell,' she says, describing informal excursions she organises to Spain and Portugal to track them down in the wild; missions that leave some locals so perplexed 'they think we're barking'.

When people want to know why she does it, she tries to explain: 'To look at them, to photograph them, to sit amongst them and be happy.'

Her inquisitors respond, 'Is that it?'

Petherbridge laughs. 'Well, actually, yeah.'

Yet things are changing. Many wild daffodils she has found in recent years are now buried under concrete and Petherbridge fears for the future of native populations. Professor Spencer Barrett is also deeply troubled. '*Narcissus* is under threat,' he says. 'This is a serious problem because many of the species in the Mediterranean are quite rare and they occur in habitats that are getting the hell hammered out of them by development.'

Barrett has witnessed 'incredible hillsides' of narcissi being obliterated by time-share apartments. 'There is a real concern about losing some of these species because the Mediterranean is under such pressure, often from Brits,' he adds with irony, 'going there for their holidays.'

Over-harvesting and habitat destruction are driving vulnerable individual populations underground; literally and, in an increasing number of instances, permanently. On a global level, the Red List of Threatened Species, compiled by the International Union for the Conservation of Nature, classifies five species as endangered, including Andalusia's *Narcissus longispathus* and the narrow-leaved *Narcissus angustiformis* from the Ukraine.

Environmentalists continue efforts to combat the problem. Ukraine's enormous 257-hectare Narcissus Valley Massif shelters more than 400 different plant species and Romania established the Negrași Daffodil Meadow, within which grows the rare *Narcissus poeticus* subsp. *radiiflorus.* In the United Kingdom some wild daffodil populations are protected within special reserves; the Gloucester Wildlife Trust, for example, conserves *Narcissus pseudonarcissus* meadows in the so-called Golden Triangle (named for the daffodils which have long grown there) of England's north-west.

In the realm of commerce and the profitable cut flower industry, a tiny fraction of daffodil varieties dominate to the exclusion of almost everything else. Bespoke heirloom bulb specialists cater for the adventurous and motivated; as American supplier Old House Gardens argues, quite apart from the beauty and character of botanical old-timers, 'the only way to preserve these living artifacts — and their incredible genetic resources — is to grow them.'

Opting to do that can generate some unexpected consequences. It has left Ron Scamp, for example, metaphorically embedded in my mother's life. The last time I visited her garden I noticed a little floral assembly of dramatic orange-yellow Division 11a split corona daffodils — a Scamp hybridising specialty. One looked suspiciously like his 2004 'Crugmeer', a 21st-century bloom so startlingly two-dimensional it could have been flattened by a vice.

In 2015 the Royal Horticultural Society awarded Scamp its Peter Barr Memorial Cup in order to honour an outstanding daffodil career that has seen him (and in the latter years his son Adrian) register around 400 new varieties. The breeder told me he intends to retire, leave his firm's 14-acres (5.6 hectares) in Falmouth in the soil-stained hands of his descendants, and go cruising, far from the flowerbeds that have made his name.

That will not stop my mother ordering new bulbs, both leading-edge and historic, while keeping a close watch on her soil in case other varieties planted long

before she inherited this landscape suddenly elect to bloom. She identified one such daffodil, a bi-coloured Division 1 with a lemony trumpet and off-white perianth as 'Moira O'Neill', although that may be wishful thinking (the surname mirrors our own) as this flower, like so many others, seems to have drifted from view since George Engleheart registered it in 1923.

Far easier to identify correctly is 'Rip van Winkle', a yellow, impish Division 4a that is definitely part of her *Narcissus* menagerie. Believed to have hailed from Ireland at some point before 1884 this nimble migrant is every inch an oddity — dwarfish in stature with wild starburst flowers that are comprised of slender but thoroughly unruly perianth segments, the tips of which resemble crochet hooks.

Why 'Rip van Winkle' is so-called remains a puzzle. Possibly it relates to the fact that this is one of the earliest flowering miniature varieties, but it may have been named after the character created by American author Washington Irving in the early nineteenth century because it, too, resurfaced so long after having been last seen that the world had effectively forgotten it.

In Britain the Daffodil Society hears from people who have moved into old houses with aged, overgrown gardens, or into villages that have long-neglected churchyards. They clear away the overgrowth and witness daffodils appearing that they have never seen before, occasionally sending photographs or little shrivelled examples to the Society, which Petherbridge and her colleagues do their best to identify. Some of these bulbs may have lain dormant — starved of light, air or water — since the early 1900s, Petherbridge says. 'Once they get what they need they will flower.'

The American Daffodil Society so often faces the question 'How long do daffodil bulbs last?' that it has an answer posted online. 'Under good growing conditions, they should outlast any of us,' it counsels. 'While some kinds of bulbs tend to dwindle and die out, daffodils should increase.'

Not all daffodils are born equal. Some, particularly among the antique varieties, exhibit more staying power than most. Jan de Graaff, a Netherlands-born member of the notable Dutch bulb dynasty (his family's involvement dates back to 1723), became a prominent player in American horticulture, and considered the modern daffodil — good as he thought it was — 'not good enough'.

'There is another quality, too, in the older daffodils — a beauty, a balance, something that I might call a personality,' he wrote in the *American Daffodil Yearbook*, 1960, as he ruminated on the ability of some narcissi, specifically the wild varieties and more ancient hybrids, to endure despite all the odds. 'They grow without care or attention,' he noted. 'They are picked, their foliage is cut or mowed along with the grass. They are trampled on, dug up and thrown out, only to grow again where they land. Obviously there is an amazing strength in those daffodils. That strength, that power to survive ...'

Dotted around the globe are amateur and professional horticulturalists hunting for such daffodils. Caroline Thomson, a descendent of the hybridising Backhouse clan, has made it her mission to find missing bulbs created by her daffodil-breeding family to complete the Backhouse Heritage Daffodil Collection at Rofsie Estate in Scotland, and to secure National Collection status. In 2015 news broke that 'Bulwark' and 'Rosslare', two other 'lost' daffodils bred by the Brodie in the 1920s, had been discovered in a specialist bulb nursery in Australia and returned 'home' to Morayshire, Scotland.

Some daffodils repay sanctuary by being under constant observation. Kew Gardens cultivates thirty-five naturally occurring *Narcissus* species and fifty-odd hybrid cultivars, two of which — *Narcissus pseudonarcissus* and 'February Gold', a Division 6 dwarf *cyclamineus* registered in 1923 by de Graaff Bros — have been part of a study of flowering dates since the 1950s. Like canaries in a coalmine their behaviour speaks of atmospheric changes as spring has soldiered forwards. In the 1980s Kew's daffodils flowered on or around 12 February every year. Twenty-five years later the average date was closer to 27 January.

Sandra Bell, Kew's Wildlife and Environment Recording Coordinator, is cautious about what these *Narcissus* flowering dates results mean, as the differences from one year to another can be dramatic. She is also, as yet, hesitant to categorically link the trend to climate change over the second half of the twentieth century. 'We will just have to wait and see what happens,' she counters, agreeing that 'climate change offers the likeliest explanation for the data we have recorded.' However, the researcher adds, 'the relationships between climate and flowering are complicated and it is not always easy to tease out cause and effect.'

Cause, effect and meaning ripple through this flower's tale — a narrative in which the past, the present and the future continually intersect. Nowhere is this more visible than in the birth of what might be called the 'digital daffodil' courtesy of DaffSeek, an online database sponsored by the American Daffodil Society containing around 29,600 *Narcissus* records, plus around 29,400 photographs by 308 photographers in twenty-two countries.

The brainchild of Nancy Tackett and Ben Blake, married Californians who jointly won an American Daffodil Society Gold Medal and the Peter Barr Memorial Cup for the initiative, it launched in 2006, runs in nine languages, and provides details of daffodils by name, classification, year of registration (where available), hybridiser, seed and pollen parents and country of origin. A team of internationally recognised botanical experts including Brian Duncan from Northern Ireland, Juan Andrés Varas Braun from Chile, Germany's Theo Sanders, and Sally Kington from England, contribute to the open access database.

Echoing the democratising effect of Carl Linnaeus's taxonomy two and half centuries ago, DaffSeek allows anybody internet-enabled to potentially identify any *Narcissus* they find. In true multi-hemisphere style, co-founder Blake worked with New Zealand software developer and fellow daffodil enthusiast Lachlan Keow to create customised features such as DaffSeek's 'Pedigree' function which reveals every known descendant of each recorded daffodil.

This may sound arcane but it lays bare the intricate web of daffodil bloodlines; a miasma of interwoven DNA and drama. Additional clues can be found within DaffLibrary, the American Daffodil Society's collection of catalogues, publications and correspondence.

There is such beauty in the detail of these historical records, and evidence that those who really look at daffodils begin to see things a different way. Take Guy L. Wilson, the Irish breeder who spelt out his reaction to marble white daffodils, writing in the *American Daffodil Yearbook*, 1937:

> They seem to revel in sunshine ... and in sunny weather their purity and substance increases day by day. The eye delights in the faint blue shadows cast by the sunlight amongst their petals, throwing into exquisitely delicate contrast the cool faint lemon, ivory, and cream tones in their crowns ... In the evening after sunset, as twilight deepens, they seem to distill about themselves a magic light of their own, and cast an enchanted spell of unearthly beauty and peace.

Another archival gem dates from the spring of 1897 when Irish bulb trader William Baylor Hartland received a letter penned in green ink, carefully bound with a matching green ribbon. The missive thanked him for a box of daffodils he had sent, a gift the writer described as those 'bonnie, gold stars of Spring', prompting Hartland to contemplate in his catalogue why it is that daffodils hold such 'special charm for pent-up city life, and where that life in this busy world of ours, has become such worry, amid the greatest struggle for existence'.

The daffodil has threaded its way through much of human history and within its story lies our own — encompassing love, lust, desire, treachery, joy and a quest for knowledge that can never be constrained. Like the eighteenth-century daffodils hybridisers we are at the cusp of a revolution that promises to redefine who we are. Computer-powered biology is allowing genetic engineering to take place on a scale previously impossible and alongside the disquieting spectres of enhanced humans and bio-weaponry comes the reality of life beyond our planetary home.

Since the year 2000, humans have continuously lived in orbit aboard the International Space Station, built by five space agencies representing fifteen countries. In August 2015 the onboard astronauts of Expedition 44 took their first bites of a space-grown vegetable — red cos (romaine) lettuce dressed with olive oil and balsamic vinegar. 'The crew just likes to nurture them,' said scientist Giola Massa in a statement that pointed out the value of space-cultivated food to projects such as a voyage to Mars, but also likened plants to pets, and explained that astronauts felt comforted by merely having them onboard.

Here on Earth the climate may be shifting but in other things constancy remains. Scientists continue to examine the daffodil. They mine it for potentially healing chemicals, refine pest-control methods to protect it, and push it ever further in an attempt to extend flowering times and the life of blooms. Though old-timers worry that interest is waning, organisations as far afield as Tasmania, France and the Ukraine hold daffodil festivals for those so enamoured, and the World Daffodil Convention, a conference scheduled to take place in the American city of St Louis in April 2016, continues to be organised every four years.

On a day-to-day level, our world keeps turning. Dramatic events, like the one I experienced when faced with illness, can force you to stop and smell the roses — or the daffodils, as it may be — but life prevails.

On the day I visited Signature Prints in Sydney to see the enigmatic designer Florence Broadhurst's secret jonquils, CEO David Lennie showed me a 'new' daffodil from a vintage New Zealand archive called Blume. Florence Broadhurst designs have captivated the world and Lennie hoped the forgotten Blume might also flourish.

He flipped through a fading wallpaper sample book and stopped at a *Narcissus* line drawing. 'It's stylistically art deco,' he told me of the deceptively uncomplicated image. 'Quite lovely. Never revived.'

Some argue that all histories are intrinsically subjective. Certainly this 'biography' of the daffodil is more selective and personal than most. At its heart lies the relationship between humans and flowers, something I was reminded of upon meeting landscape designer Phillip Johnson, who in 2013 led the first Australian team to win Best in Show at the Royal Horticultural Society's Chelsea Flower Show.

Johnson adores nature. His particular passion is the 21st-century issue of sustainability, and as he creates native gardens in urban Australian environments he finds little use for the daffodil. Yet he tells the story of an aspidistra his great-grandparents brought with them from England when they migrated to Australia long ago. Generations later his parents created an English-style garden in Victoria where every aspidistra is descended from that original plant.

I myself have been transplanted, and when I come across a daffodil in Australia I feel the comfort of connection. Yet to see narcissi flowering in so many forms and colours across my mother's garden is something else again. Here they represent the past and the future, the capacity for renewal and for healing; the fact that darkness arrives and settles but, like winter, can be overcome. At first glance nothing might appear less important than a daffodil yet with every spring they captivate me anew.

As I was researching this book I happened to chat to a friend who discovered as a child that *Narcissus* is poisonous the painful way after attempting to eat one. For several hours he seemed just fine. Then he was violently ill.

'Has the daffodil changed the world?' he asked me.

I had to think about that for a moment. Then I found myself saying, with an impulse that surprised me, 'My world? Yes indeed.'

Acknowledgments

I owe a debt of gratitude to everybody who took the time to converse with me, either verbally and electronically — a list of luminaries that includes John Bailey, Professor Spencer Barrett, Linda and Sara van Beck, Sandra Bell, Dr Peter Brandham, Professor Richard Clough, Jan Dalton, Lyn Edmonds, Dr Linda Evans, Professor Christopher Eyre, Dr Benjamin Henry, Professor Harold Koopowitz, Jackie Petherbridge, Professor Diego Rivera, Rupert Sanderson, Ron Scamp, Dr Robert Scotland, Kevin Stephens, Nancy Tackett, Margaret Woodward, Melanie Underwood, Timothy Walker, Richard Wilford and Joy Uings. Apologies to anybody I have missed.

Authors, just like daffodil hybridisers, find themselves indebted to the work that has gone before, sometimes centuries earlier, and to its custodians such as the American Daffodil Society's Daffseek.org, the UK's Daffodil Society and the Royal Horticultural Society.

To the team at HarperCollins Books, especially to my publisher Catherine Milne, thank you for believing in this book.

Thank you to my family for always supporting me — in particular my brother, Patrick O'Neill, for taking so many beautiful photographs of these daffodils and to my mother, Judith O'Neill, for drawing them and loving them — and us — so well.

ROYAL HORTICULTURAL SOCIETY
DAFFODIL CLASSIFICATION

Whether of wild or cultivated origin, once a selection has been distinguished by a cultivar name it should be assigned to Divisions 1–12. Daffodils distinguished solely by botanical name should be assigned to Division 13.

NOTES

1. The characteristics for Divisions 5 to 10 are given for guidance only; they are not all necessarily expected to be present in every cultivar assigned to those divisions.
2. Divisions 12 and 13 are not illustrated owing to the wide variation in shape and size between the flowers involved.

DIVISION 1 – TRUMPET DAFFODIL CULTIVARS
One flower to a stem; corona ('trumpet') as long as, or longer than the perianth segments ('petals').

DIVISION 2 – LARGE-CUPPED DAFFODIL CULTIVARS
One flower to a stem; corona ('cup') more than one-third, but less than equal to the length of the perianth segments ('petals').

DIVISION 3 – SMALL-CUPPED DAFFODIL CULTIVARS

One flower to a stem; corona ('cup') not more than one-third the length of the perianth segments ('petals').

DIVISION 4 – DOUBLE DAFFODIL CULTIVARS

One or more flowers to a stem, with doubling of the perianth segments or the corona or both.

DIVISION 5 – TRIANDRUS DAFFODIL CULTIVARS

Characteristics of *N. triandrus* clearly evident: usually two or more pendent flowers to a stem; perianth segments reflexed.

DIVISION 6 – CYCLAMINEUS DAFFODIL CULTIVARS

Characteristics of *N. cyclamineus* clearly evident: one flower to a stem; perianth segments significantly reflexed; flower at an acute angle to the stem, with a very short pedicel ('neck').

DIVISION 7 – JONQUILLA AND APODANTHUS DAFFODIL CULTIVARS

Characteristics of Sections Jonquilla or Apodanthi clearly evident: one to five (rarely eight) flowers to a stem; perianth segments spreading or reflexed; corona cup-shaped, funnel-shaped or flared, usually wider than long; flowers usually fragrant.

DIVISION 8 – TAZETTA DAFFODIL CULTIVARS

Characteristics of Section Tazettae clearly evident: usually three to twenty flowers to a stout stem; perianth segments spreading not reflexed; flowers usually fragrant.

DIVISION 9 – POETICUS DAFFODIL CULTIVARS

Characteristics of N. *poeticus* and related species clearly evident; perianth segments pure white; corona very short or disc-shaped, not more than one-fifth the length of the perianth segments; corona usually with a green and/or yellow centre and red rim, but sometimes wholly or partly of other colours; anthers usually set at two distinct levels; flowers fragrant.

DIVISION 10 – BULBOCODIUM DAFFODIL CULTIVARS

Characteristics of Section Bulbocodium clearly evident: usually one flower to a stem; perianth segments insignificant compared with the dominant corona; anthers dorsifixed (ie attached more or less centrally to the filament); filament and style usually curved.

DIVISION 11 – SPLIT-CORONA DAFFODIL CULTIVARS

Corona split – usually for more than half its length
a) Collar Daffodils: Split-corona daffodils with the corona segments opposite the perianth segments; the corona segments usually in two whorls of three.

b) Papillon Daffodils: Split-corona daffodils with the corona segments alternate to the perianth segments; the corona segments usually in a single whorl of six.

DIVISION 12 – OTHER DAFFODIL CULTIVARS

Daffodil cultivars which do not fit the definition of any other division.

DIVISION 13 – DAFFODILS DISTINGUISHED SOLELY BY BOTANICAL NAME

SECTION TAPEINANTHUS

Autumn flowering; one to four flowers to a rounded stem; leaves very narrow, glaucous, not always present on flowering bulbs; flower ascending, yellow; corona absent or rudimentary; anthers widely exserted from the tube, much shorter than the filaments, dorsifixed.

SECTION SEROTINI

Autumn flowering; usually one to two flowers to a rounded stem; leaves very narrow, glaucous, not always present on flowering bulbs; perianth segments pure white, usually twisted; corona very short, yellow, orange or green; anthers included in or slightly exserted from the tube, longer than the filaments, dorsifixed; flowers fragrant.

SECTION AURELIA

Autumn flowering; three to twelve flowers to a compressed stem; leaves flat not channelled, glaucous; flowers white; corona rudimentary or absent; filaments unequal in length; anthers exserted from the tube, dorsifixed; flowers fragrant.

SECTION TAZETTAE

Autumn to spring flowering; three (rarely two) to twenty flowers to a usually compressed stem; leaves flat or channelled, usually glaucous; flowers white, yellow or bicoloured; anthers included in or slightly exserted from the tube, much longer than the filaments, dorsifixed; flowers fragrant. The rounded stem and green leaves of N. *aureus* atypical, also the orange corona of N. *elegans*.

SECTION NARCISSUS

Spring flowering; usually one flower (exceptionally two to four) to a compressed stem; leaves flat not channelled, glaucous; perianth segments pure white; corona disc-shaped or very shallow, sometimes of a single colour, but usually with base green, mid-zone yellow and rim red or orange and often scarious; anthers partly exserted from the tube, much longer than the filaments, dorsifixed; flowers fragrant. Section covers N. *poeticus*.

SECTION JONQUILLA

Spring flowering; one to five (rarely eight) flowers to a rounded stem; leaves narrow or semi-cylindrical, green; flowers yellow, never white; perianth segments spreading or reflexed; corona usually cup-shaped, usually wider than long; anthers included in or partly exserted from the tube, much longer than the filaments, dorsifixed; flowers fragrant. The autumn flowering, green-flowered N. *viridiflorus* is atypical.

SECTION APODANTHI

Spring flowering; one flower or two to five to a somewhat compressed stem; leaves narrow, channelled, glaucous; flowers white or yellow, never bicoloured; perianth segments spreading or slightly reflexed; corona cup-shaped, funnel-shaped or flared, usually wider than long; anthers included in the tube or three included and three exserted, much longer than the filaments, dorsifixed.

SECTION GANYMEDES

Spring flowering; one flower or two to six to an elliptical or cylindrical stem; flowers pendent, white or yellow or somewhat bicoloured; leaves flat or semi-cylindrical; perianth segments reflexed; corona cup-shaped (rarely campanulate); anthers three included in the tube, three exserted (often beyond the corona), equal to or much shorter than the filaments, dorsifixed. Section covers *N. triandrus*.

SECTION BULBOCODIUM

Autumn to spring flowering; one flower to a rounded stem; leaves narrow, semi-cylindrical; flowers white or yellow; perianth segments insignificant compared with the dominant corona; anthers widely exserted from the tube (often beyond the corona), much shorter than the filaments (which are usually curved), dorsifixed.

SECTION PSEUDONARCISSUS

Spring flowering; usually one flower to a more or less compressed or sometimes rounded stem; leaves flat or channelled, usually glaucous; flowers white, yellow or bicoloured; perianth segments usually spreading or inflexed; corona more or less cylindrical, often flared at mouth, yellow or white (never orange or red); anthers exserted from the tube, equal to or shorter than the filaments, sub-basifixed. The green leaves, rounded stem and strongly reflexed perianth segments of *N. cyclamineus* and the two to four flowers to a stem of some species including *N. alcaracensis*, *N. longispathus* and *N. nevadensis* are atypical.

NOTE: Hybrids distinguished solely by botanical name are also assigned to this Division. In the Register, daffodils distinguished solely by botanical name are listed separately from those with cultivar or group names.

Text & images © RHS 2012

Bibliography / Further Reading

American Daffodil Society, *1966 Daffodil Handbook*, DaffLibrary, American Daffodil Society, Inc., available at dafflibrary.org.

Anon, 'Mr Robert Sydenham', *Edgebastonia,* vol XX, no 229, June 1900 pp 101–108, DaffLibrary, American Daffodil Society, Inc., available at dafflibrary.org.

Anon, 'The Daffodil King. Interview with Mr Peter Barr', *Bendigo Advertiser*, Friday 23 November 1900, p 3.

Anon, 'The Daffodil King', *Tuapeka Times,* 6 November 1909, p 4.

Anon, 'The Englishman of the Narcissi', *El Correo Gallego,* 14 December 1888, DaffLibrary, American Daffodil Society, Inc., available at dafflibrary.org.

Baker, Gilbert John, 'Review of the Genus Narcissus', *The Gardeners' Chronicle and Agricultural Gazette,* 17 April 1869, 8 May 1869, pp 416–417, 529.

Barr, Peter, *Barr's Daffodils: Barr's Descriptive Catalogue of Hardy Daffodils,* Barr & Son, Autumn 1886, DaffLibrary, American Daffodil Society, Inc., available at dafflibrary.org.

Barr, Peter, *Reading on the Cultivation of the Daffodil,* delivered at the monthly meeting of the Sea Point Horticultural Society, Cape Town: J.C. Juta & Co., 1901), accessed online via State Library Victoria www.slv.vic.gov.au

Barr, Peter, *Peter Barr's Travels* (1887, 1892), DaffLibrary, American Daffodil Society, Inc., available at dafflibrary.org.

Barr, Peter, 'Raising New Daffodils', *West Gippsland Gazette*, 2 October 1900, p 4.

Barr, Peter & Burbidge, F.W., *Ye Narcissus or Daffodyl Flowre, and hys Roots,* Barre & Sonne, pamphlet, 1884, DaffLibrary, American Daffodil Society, Inc., available at dafflibrary.org.

Beale, Catherine, 'A Notable Narcissus Nursery', *Hortus* 81, Spring 2007.

Berra, Tim M., Alvarez, Gonzalo & Ceballos, Francisco C., 'Was the Darwin/Wedgwood Dynasty Adversely Affected by Consanguinity?', *BioScience* 60(5): pp 376–383, 2010.

Blanchard, John, *Narcissus A Guide to Wild Daffodils*, Alpine Garden Society, 1990.

Bowles, E.A., *A Handbook of Narcissus,* Martin Hopkinson Ltd, 1934.

Birchfield, Mrs James, 'All Set for the Daffodil Season', *The Daffodil Bulletin,* February 1962, pp 1–2, DaffLibrary, American Daffodil Society, Inc., available at dafflibrary.org.

Bourne, Rev. S. Eugene, *The Book of the Daffodil,* John Lane: The Bodley Head, 1903.

Brady, Rose, 'The "King Alfred" Daffodil Story', *The Daffodil Society Newsletter,* 1999.

Brockbank, W., 'The Daffodil "Sir Watkin"', *The Gardeners' Chronicle,* 29 December 1894, pp 773–774.

Brockbank, William, 'Edward Leeds', *The Gardener's Chronicle*, November 10 and November 14, 1894, pp 561–2 and 625–6.

Brodie, Montague Ninian Alexander, 'Brodie of Brodie talks to Sam Marshall', Moray Firth Radio interview, undated, held at Ambaile, www.ambaile.org.uk.

Burbidge, F.W., *The Narcissus: Its History and Culture,* L. Reeve & Co., 1875.

Burbidge, F.W., *Letters from Burbidge, Engleheart and Others* (1886–1905), DaffLibrary, American Daffodil Society, Inc., available at dafflibrary.org.

Calvert, Albert F., *Daffodil Growing for Pleasure and Profit,* Dulau & Co. Ltd., 1929.

Canadian Cancer Society *Very Special People: The Achievements of the Canadian Cancer Society in Ontario,* Toronto: Canadian Cancer Society, Ontario Division, 1984.

Cartwright, R. Chatwin & Goodwin, Arthur R., *The Latest Hobby: How to Raise Daffodils from Seed,* Cartwright & Goodwin pamphlet, 1908.

Curthoys, M.C., 'Barr, Peter (1826–1909)', *Oxford Dictionary of National Biography,* Oxford University Press, May 2009, www.oxforddnb.com/view/article/96755

Dalton, Jan, 'The English Lent Lily', *Daffodil Society Journal,* 2012.

Dalton, Jan, *The Daffodil Society – A Potted History,* The Daffodil Society.

Darwin, Charles, *On the Origin of Species by Means of Natural Selection, or the Preservation of Favoured Races in the Struggle for Life.,* London: John Murray, 1859.

Darwin, Charles, *The Correspondence of Charles Darwin, Volume II: 1863,* Cambridge University Press, 1999.

Elliott, Brent, 'The Victorian Language of Flowers', *Plant-Lore Studies,* London: The Folklore Society, University College of London, 1984, pp 61–65.

Ellwanger, George H., *The Garden's Story,* New York: D. Appleton and Company, 1889.

Engleheart, Rev G.H., 'Seedling Daffodils', *Journal of the Royal Horticultural Society,* vol XI, 1889, pp 93–103.

Gebbie, James, 'A Noted Visitor', *Otago Witness,* 3 May 1900, p 8., letter to the editor from James Gebbie, Public Gardens, Oamaru.

Graaff, Jan de, 'Daffodils – A Review and Preview', The 1960 American Daffodil Yearbook, American Daffodil Society, 1960, pp 10–20, DaffLibrary, American Daffodil Society, Inc., available at dafflibrary.org.

Gray, Alec, 'Small Hybrid Daffodils', *Journal of the Royal Horticultural Society,* 1965, pp 372–381, DaffLibrary, American Daffodil Society, Inc., available at dafflibrary.org.

Grimshaw, John, 'Leonardo da Vinci, Painter at the Court of Milan, John Grimshaw's Garden Diary blog, 7 January 2012.

Hartland, Wm Baylor, *Hartland's Conference Daffodils, (*Wm Baylor Hartland's, 1897) DaffLibrary, American Daffodil Society, Inc., available at dafflibrary.org.

Hartland, Wm Baylor, *Ye Original Little Booke of Daffodils,* (Wm Baylor Hartland's, 1885) DaffLibrary, American Daffodil Society, Inc., available at dafflibrary.org.

Haworth, A.H. Esq., 'A New Arrangement of the Genus Narcissus', *Transactions of the Linnean Society of London*, vol 5, issue 1, February 1800, pp 242–245.

Herbert, William *Amaryllidaceae: Preceded by an Attempt to Arrange the Monocotyledonous Orders and Followed by a Treatise on Cross-bred Vegetables and Supplement,* James Ridgeway, 1837.

Herbert, Rev. William, 'On the Production of Hybrid Vegetables', *Transactions of the Horticultural Society of London, vol 4*, 1822, pp 15–50.

Herbert, William, *Supplement to the Works of the Hon and Very Rev William Herbert, Dean of Manchester,* London: H.G. Bohn, 1846.

Jacob, Joseph, *Daffodils with Eight Coloured Plates,* London, J.C. & E.C Jack, 1910.

Johnson, Philip, *Connected – The Sustainable Landscapes of Phillip Johnson,* Murdoch Books, 2014.

King, Mrs Francis, *Chronicles of the Garden,* Charles Scribner's Sons, 1925.

Kingsbury, Noel, 'A Gothic Hunt for Heirloom Daffodils', *Noel's Garden Blog',* 28 April 2012.

Kingsbury, Noel. *Daffodil: The Remarkable Story of the World's Most Popular Spring Flower*, Timber Press, 2013.

Kirby, A.M., *Daffodils – Narcissus and How To Grow Them,* London: William Heinemann, 1907.

Martin, A.M. *'The Perils of Plant Collecting'*, Medical Historian – The Bulletin of the Liverpool Medical History Society, no 16, 2004–2005 session, pp 27–42.

National Trust for Scotland, *Ian Brodie: A Chieftain in the World of Daffodils*, Scotland: National Trust for Scotland, 1999.

Nevill, Dorothy, *The Life and Letters of Lady Dorothy Nevill*, Methuen & Co Ltd, 1919.

Nuñez, Diego Rivera, Castro, Concepción Obón de, Ruíz, Segundo Ríos, Ariza, Francisco Alcaraz, 'The Origin of Cultivation and Wild Ancestors of Daffodils (*Narcissus* subgenus *Ajax*) (Amaryllidaceae) from an Analysis of Early Illustrations', *Scientia Horticulturae*, vol 98 no.4, 2003, pp 307–330.

Parkinson, Anna, *Nature's Alchemist: John Parkinson, Herbalist to Charles I*, Lincoln Publishers, 2007.

Parkinson, John, *Paradisi in Sole Paradisus Terrestris. Faithfully reprinted from the edition of 1629.*, Methuen & Co., 1904.

Pizzorusso, Ann C., *Tweeting Da Vinci*, Da Vinci Press, 2014.

Robinson, William, *The Wild Garden*, London: John Murray, 1870.

Royal Horticultural Society *1913 Daffodil Yearbook*, Royal Horticultural Society, 1913.

Sala, George Augustus, *Twice Round the Clock, or The Hours of the Day and Night in London*, London: Houlston and Wright, 1859.

Spotts, Bob, 'The Quest for Beautiful Green Daffodils', *Bulletin* of the Horticultural Society of Canberra, 2013, DaffLibrary, American Daffodil Society, Inc., available at dafflibrary.org.

Tait, Alfred W., *Notes on the Narcissi of Portugal,* Pôrto, 1886, DaffLibrary, American Daffodil Society, Inc., available at dafflibrary.org.

Throckmorton, Tom D., 'Maybe It's Time For Recess', *The Daffodil Journal,* vol. 16, no.2, December 1979, p 92, DaffLibrary, American Daffodil Society, Inc., available at dafflibrary.org.

Tompsett, Andrew, *Golden Harvest. The Story of Daffodil Growing in Cornwall and the Isles of Scilly,* Alison Hodge Publishers, 2006.

Walther, Angela Jean, *From Fields of Labor to Fields of Science: The Working Class Poet in the Nineteenth Century,* Iowa State University, Graduate Theses and Dissertations, Paper 12503. 2012.

Williams, F.J., 'J.C. Williams – An Enthusiast', Presidential Address to Royal Institute of Cornwall, November 1998.

Willis, David, *Yellow Fever: A prospect of the History and Culture of Daffodils,* David Willis, 2012) DaffLibrary, American Daffodil Society, Inc., available at dafflibrary.org.

Wilson, Guy L., 'Some Modern Daffodils for Garden Decoration', *The American Daffodil Yearbook,* The American Horticultural Society, 1937, pp 14–22, DaffLibrary, American Daffodil Society, Inc., available at dafflibrary.org.

Wolley-Dod, Charles et al, *Extracts from Correspondence* (1885–1891), DaffLibrary, American Daffodil Society, Inc., available at dafflibrary.org.

Wordsworth, Dorothy, *Journals of Dorothy Wordsworth: The Alfoxden Journal 1798, The Grasmere Journals 1800–1803,* ed. Mary Moorman, Oxford University Press, 1971.

Uings, Joy, *Edward Leeds: A Nineteenth Century Plantsman*, dissertation submitted to the University of Manchester for the Certificate of Continuing Education in Garden and Landscape History, 2003.

Yerger, Meg, 'The Brodie – Champion of Poets', *The Daffodil Journal,* vol 16, no 4, June 1980, pp 220–223, DaffLibrary, American Daffodil Society, Inc., available at dafflibrary.org.

Zandbergen, Matthew, 'How Ramsbottom Gave New Life to the Narcissus', Sassenheim, Holland, DaffLibrary, American Daffodil Society, Inc., available at dafflibrary.org.

INDEX

IMAGE CAPTIONS

2-3
Narcissus Tazetta, one of the oldest daffodil varieties and most widely travelled, as drawn by the German botanical artist Georg Dionysius Ehret (1708–1770).

6
A *Narcissus* eye-view of my parents' home, and my mother's garden, at the height of daffodil season with 'Emperor' standing regally in the foreground. When it first appeared in 1869 this Division 1 trumpet, the progeny of pioneering English breeder William Backhouse, was considered one of the finest specimens ever seen.

10-11
Daffodils fracturing England's wintery grassland at The Stray, in Harrogate, North Yorkshire.

14-15
Japanese artist Konan Tanigami (1879–1928) pioneered the depiction of western flowers with the use of traditional woodblocks. This daffodil composition is from his Seiyou Sokazufu (Pictorial Album of Western Plants and Flowers: Spring).

22-23
Poetry in motion, and in bloom, at Ullswater in England's Lake District. Here, in 1802, Dorothy Wordsworth and her brother William unexpectedly discovered a 'host of golden daffodils'.

27
Narcissus, fatally trapped by his own reflection, as depicted in a detail of 'Echo and Narcissus', a 1903 oil painting by the British artist John William Waterhouse (1849–1917).

30-31
The face of 'Fortune'. This glorious flower first appeared in the garden of Tottenham plantsman Walter Ware in 1916, was brought to market by Scotland's Brodie of Brodie, and took the daffodil world by surprise.

34
John Parkinson wove his love of the daffodil through the pages of his ground-breaking 1629 book *Paradisi in Sole Paradisus*. Reproduced here is the title page.

38-39
The daffodil makes a stylised appearance (lower left) in the Helmingham Herbal and Bestiary, a rare manuscript created in England circa 1500 by an unknown author.

44
William Herbert, known as 'The Dean' to daffodil aficionados, as he looked in 1795 (as an Eton schoolboy of seventeen) to royal portrait painter Sir William Beechey.

50-51
Daffodils represented both romance and potential to the Isles of Scilly and beyond, as this vibrant example of 1920s publicity material from the Empire Marketing Board shows.

55
The intimacy of *Narcissus,* as captured in 'Woman at Tea Time' by contemporary American artist Daniel F Gerhartz.

59
'Daffodils in Blue and White Jug', a modern-day watercolour on paper by Sally Maltby.

65
'Daffodil with Red Admiral', a 1568 watercolour by Jacques Le Moyne de Morgues (1533–1588), the multi-talented French artist and explorer.

67
Peter Barr, the entrepreneurial, Scottish-born plantsman who became known across the world as the 'Daffodil King'.

72–73
Impoverished street-sellers trying to earn a crust by selling cut daffodils to the rich in London's Piccadilly, recorded in the spring of 1887 by oil painter Edward Clegg Wilkinson.

77
The 'Large double Daffodil', identified as N. Pseudo-Narcissus var. Grandiplenus by F.W. Burbidge in his 1875 book *The Narcissus Its History and Culture.* Burbidge described it as 'beautiful', 'monstrous', and the first daffodil known to have been grown from seed, by John Parkinson in 1618.

89
Daffodil retail merged art with marketing, as this 1892 catalogue

from New York's Peter Henderson & Company shows. The woman in the white dress represents Henderson's ideal customer, somebody with money, taste and a love of the exciting new Narcissi.

93

Double daffodils, categorized as Division 4, possess an eerie beauty. They fascinated early botanists such as John Parkinson, and continue to intrigue today.

100–101

Buzzing between *Narcissus poeticus* blooms, a pollinating bee is frozen in action by a seventeenth-century botanical artist.

105

Wild daffodils have long been taken for granted as part of the English landscape, as this illustrated plate from the 1881 book *At Home* by J G Sowerby and Thos. Crane shows.

109

My mother has repeatedly planted unidentified mixed packs of *Narcissus* bulbs for her garden and seldom

knows what will flower next. Division 4 doubles such as this pink and white beauty could well be 'Spaniards Inn', registered in 1993 by Dan du Plessis and bred by Brian S. Duncan.

113

Gustav Klimt (1862–1918) created 'Dancer' in 1916, posing his bare-breasted model with golden daffodils to create an image of iconic eroticism.

116–117

'Jenny', a Division 6 dwarf daffodil with swept-back petals created pre-1943 by English breeder Cyril F. Coleman (1898–1980) who specialized in *Cyclamineus*.

121

This charming nineteenth-century Chinese watercolour of daffodils demonstrates style rather than botanical accuracy.

125

Hand-coloured engraving of bunch-flowered *Tazetta Narcissi* from the epic, 1613 flower catalogue *Hortus Eystettensis* by Basil Besler (1561–1629).

128–129
Drifts of *Narcissus* create a 'Daffodil Hill' amidst the semi-wild woodlands of England's Lake District.

133
The enigmatic designer Florence Broadhurst hid bunches of jonquils within her fantastical 'Carnation' pattern so effectively that few, including at least one of her hands-on artists, had any idea they were there.

137
My mother is particularly fond of her early twentieth-century 'heirloom' daffodils. These, photographed by my brother, have been cautiously identified as 'Mrs R.O. Backhouse' (a Division 2 raised before 1921 by Sarah Backhouse) and 'Seraglio', a Division 3 daffodil from The Brodie of Brodie dating to 1926.

140–141
The Empire Marketing Board's 1920s marketing campaign urged consumers to 'Buy Empire', and depicted daffodil farming as a blissfully romantic pursuit.

145
Close up of 'Rip Van Winkle', a Division 4 yellow double daffodil bursting with promise.

149
The 'Daffodil Fairy' from *Flower Fairies of the Spring*, a classic children's book first published by English author/illustrator Cicely Mary Barker (1895–1973) in 1923.

153
Division 13's species and wild variants includes *Narcissus x odorus*, identified by Carl Linnaeus in 1756 and drawn here circa 1800 by Austrian botanical artist Franz Bauer.

156–157
American artist Joyce Gaffin called this still life watercolour of cut *Narcissi* in a blue vase simply 'Daffodils'.

161
Mural in memory of Poland's Marek Edelman, survivor of the Warsaw Ghetto Uprising, at 9b Nowolipki Street in Warsaw.

165
In 'April', commissioned as part of an 1896 calendar by the French department store La Belle Jardinière, Swiss artist Eugène Grasset (1845–1917) blended the daffodil into Art Nouveau's stylistic lexicon.

168–169
Even 'Rip van Winkle' is vulnerable to the ravages of time. This Division 4 dwarf double daffodil dates to 1884 and is considered Irish. Its raiser is unknown.

172–173
Narcissus bulbocodium growing at Cape St Vincent in Portugal alongside *Carpobrotus edulis*, the 'Hottentot Fig', an invasive plant from South Africa. Oxford University's Dr Robert Scotland takes students to this region to study this shapely, dwarf Division 13 *Narcissi*, also known as the 'Hoop Petticoat Daffodil'.

177
Dr Robert Scotland investigation of the daffodil's mystifying floral architecture lead him to review the structure of *Hymenocallis festalis*, an elegant flower with staminal filaments attached to the apex of its corona. *Hymenocallis festalis*, the so-called 'Peruvian daffodil', belongs to the Amaryllidaceae family but is not a *Narcissus*.

180–181
A hand-coloured, glass lantern slide recording an impenetrable ocean of daffodils flowing across 'Lob's Wood', part of a private Ohio estate, circa 1920.

184–185
'Jetfire', exquisitely dissected to reveal its internal architecture. Oregon-based Grant E. Mitsch (1907–1989) first registered this Division 6 dwarf Cyclamineus in 1966.

193
Burbidge drew four varieties of *N. Poeticus* (*Recurvus, Tripodalis, Stellaris, Aurantius*) for his 1875 book *The Narcissus Its History and Culture,* warning of *poeticus:* 'the odour … in close rooms, has proved extremely disagreeable, if not actually injurious to delicate persons'.

196–197
'Flowers Streaked with Gold', a vivid, 1991, daffodil-inspired oil painting by Dorset artist Philip Sutton.

201
American illustrator Frederick Richardson (1882–1937) invented 'Daffy-down-dilly' for the 1915 children's classic *Mother Goose*.

204
An unusual take on 'Itzim', a 1982 Division 6 Cyclamineus created by Oregon-based Grant E. Mitsch (1907-1989), one of America's finest daffodil breeders.

205
A flowering stiletto fashioned by London-based luxury shoe designer Rupert Sanderson to help launch his 2013 range.

208–209
Crop of 'Carlton' daffodils bunched and ready for packing. Photographed in March, in Lincolnshire, UK.

212–213
'When Flowers Return', dating to about 1911 from Netherlands-born oil painter Sir Lawrence Alma-Tadema (1836–1912).

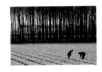

216–217
Quality-control inspectors scour the Dutch daffodil fields of Bollenstreek in the hunt for viruses that may damage the valuable crops.

222
Spring brings a kaleidoscopic array of daffodils to my mother's garden. Here are a few of the varieties that bloom, as drawn by her.

230-231
Spring brings a kaleidoscopic array of daffodils to my mother's garden. Here are a few of the varieties that bloom, as drawn by her.

254-255
April daffodils running riot across the grounds of Castle Howard in North Yorkshire.

PICTURE CREDITS